What People Are Saying About Lymph-Biologics™ and Dr. Loretta's Holistic Approach to Lymphatic Health…

"I found Dr. Loretta at a difficult, painful time in my life. She responded to me almost immediately after I requested an appointment with her. From then on, it was an absolute pleasure to work with her. While she helps reduce the pain or works on a problem in your appointment, she will also work with you to get to the root of the problem so that it is solved for good. Beyond her expertise, she is genuinely a wonderful human being. I would recommend her to anyone in the NYC/NJ area."

—Danielle Tomczak

"The directional non-force technique that Dr. Friedman uses with great care and purpose allowed me to leave the office feeling less pain, and confident in my choice to move forward with the plan of treatment. I highly recommend Dr. Friedman. She is caring, knowledgeable, and very accommodating."

—Cynthia Moss

"Dr. Friedman is honestly incredible. I think the appropriate term for her might be a 'godsend.' She has a vast knowledge of the human body, and her ability as a healer is one of a kind. You will leave her office feeling like a brand-new person, both mentally and physically. On top of all this, she sports a great sense of humor to go along with her top-notch professionalism and caring. She seems to get genuine joy from helping others."

—Zack Green

LYMPH-LINK

SOLVING THE MYSTERIES OF INFLAMMATION, TOXICITY, AND BREAST HEALTH ISSUES

DR. LORETTA T. FRIEDMAN

A POST HILL PRESS BOOK
ISBN: 978-1-63758-313-5
ISBN (eBook): 978-1-63758-314-2

Lymph-Link:
Solving the Mysteries of Inflammation, Toxicity, and Breast Health Issues
© 2022 by Dr. Loretta T. Friedman, RN, MS, DC, CCN, CNS, DACBN, DCBCN
All Rights Reserved

This book contains advice and information relating to health care. It should be used to supplement rather than replace the advice of your doctor or another trained health professional. You are advised to consult your health professional with regard to matters related to your health, and in particular regarding matters that may require diagnosis or medical attention. All efforts have been made to assure the accuracy of the information in this book as of the date of publication. The publisher and the author disclaim liability for any medical outcomes that may occur as a result of applying the methods suggested in this book.

No part of this book may be reproduced, stored in a retrieval system, or transmitted by any means without the written permission of the author and publisher.

Post Hill Press
New York • Nashville
posthillpress.com

Published in the United States of America
1 2 3 4 5 6 7 8 9 10

CONTENTS

Foreword vii
Introduction: Putting Out the Fire ix

Chapter 1   Lymph—The Dirty Little Secret Your Doctor
             Will Never Talk About. 1
Chapter 2   How the Lymph Link Will Help You Find Relief . . . 17
Chapter 3   Taking the Driver's Seat to Finding Relief 24
Chapter 4   How I Cut through the Bullsh**—and You Can Too. . .32
Chapter 5   How Toxic Are You? 40
Chapter 6   Simple Steps to Detox 55
Chapter 7   Lymph-Biologics™ and Metabolic Detox: The
             Winning Combo that Works. 61
Chapter 8   What Your Cells Can Tell You That No Doctor Will. . .76
Chapter 9   It Takes Two: Your Breasts and the Lymph Link 91
Chapter 10   The Lymph Link and Your Diet 110

Afterword 119
Glossary 121
Metabolic Detoxification Recipes
  That Will Make Your Mouth Water. 133
Appendix: What You Need to Know 201

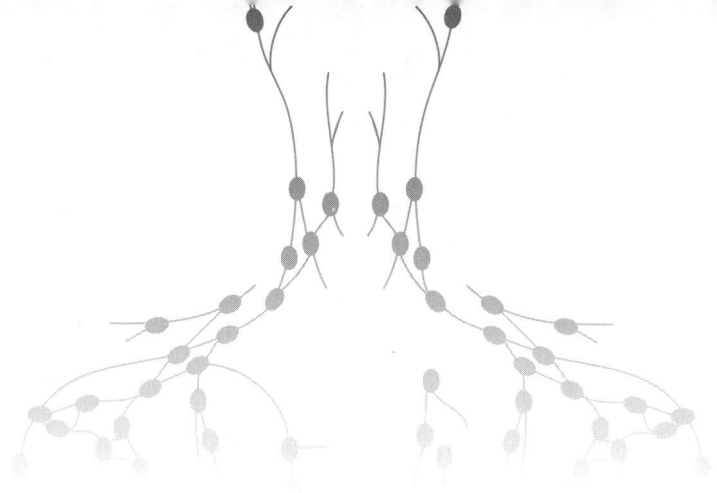

FOREWORD

Food is medicine. After all, we are what we eat. When I met Dr. Loretta fifteen years ago for some minor back pain, I knew she was a kindred spirit who believed this too.

Besides both eating healthy, we think similarly about many other things. We are both committed to top-quality service. Whether people dine at one of my restaurants or visit Dr. Loretta's office on Fifth Avenue, our job is to make people feel better. Who knew that a chef from rural France and a Manhattan lymph specialist could have so much in common?

Growing up, meat and fish were expensive, so my family ate many vegetables at all our meals. Today, in my restaurants, I still believe in these types of dishes. (Because, let's face it, no one needs to eat a fifteen-ounce steak!)

But Dr. Loretta taught me there is much more to good health than just eating well. The work I do is intense, keeping me on my feet for hours every day. Between creating in the kitchen and traveling internationally, my body can easily break down if I don't take care of it. Dr. Loretta showed me through my cell test results that, despite my clean diet, I still had inflammation and toxicity in my body.

Now, doing Lymph-Biologics™ every week and following Dr. Loretta's regimen of supplements is part of taking care of myself. In fact, my whole family—the whole *mishpucha*, Dr. Loretta would say—trusts our health to her care. Just as I weigh every ingredient I use and write my recipes down very precisely—that's the German in me—I pay very close attention to everything she tells me to do. If she tells me I'm low on vitamin D, I boost up my vitamin D. I don't wait around until symptoms start.

People are shocked when they discover my age (sixty-five at the writing of this book). Dr. Loretta always jokes with me and says, "You see how good you look, JG? That's all because of me."

It's true—between good food and Dr. Loretta, I'm set for life.

If you read this book and follow her advice, I guarantee that you will be too.

—Jean-Georges Vongerichten

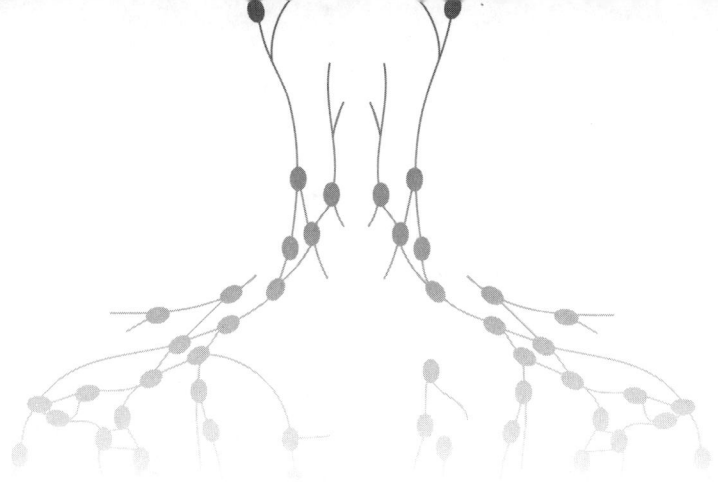

INTRODUCTION: PUTTING OUT THE FIRE

We are a country of breast disease. Lumpy, bumpy, and tender breasts are considered a normal part of a woman's menstrual cycle. Teenage girls who get their periods for the first time become part of a collective complaining committee about cramps and nausea. These uncomfortable symptoms, we are told by our mothers and healthcare professionals, are normal and simply come with the territory of being a woman.

Not so. The fact is that this pain and swelling are signs of toxicity in your body. What do I mean by toxicity? A toxin is medically defined as a poison—any substance that can cause structural or functional disturbance when taken into the body. Toxicity can come from any number of elements, including a hormonal imbalance, leaky gut syndrome, heavy metals in the bloodstream, dietary sensitivities, or chemicals in your food. This information is not the stuff they teach you in high school health class. Nor is it what most western physicians learn in medical school, where they're taught to look at human anatomy through a very small lens. During my many years as a surgical nurse, I was trained this way too. Even though I developed an excellent grasp of physiology, I hadn't yet learned to step back the way I do now and take a holistic approach to patients' health.

We can't afford not to see the forest for the trees any longer—because that forest, women's health, is on fire. The number of cases of breast- and lymph-related diseases alone in women between the ages of thirty-five and sixty is alarming. If you include the vast number of deadly diseases like asthma, type 2 diabetes, lymphoma, and breast cancer, then we've got a national crisis in women's health on our hands. Breast cancer alone kills about 43,600 people per year and will affect about one in every eight women in the US—making it the most common cancer among women. What do all of these conditions and diseases have in common? They are all linked to inflammation caused by toxicity. According to any medical dictionary, inflammation is typically defined as "a localized protective response elicited by injury or destruction of tissues, which serves to destroy, dilute, or wall off both the injurious agent and the injured tissue." Naturally, many kinds of harmful toxins could lead to that protective inflammatory response.

In 2017, Silent Spring Institute, a forerunner in women's health research, published the most thorough review of epidemiological research on breast cancer to date, indisputably linking toxicity levels to the disease. The Institute identified 296 chemicals found in common products such as hair dye and pesticides, which can lead to increases in levels of hormones known to contribute to breast cancer. Fully 219 of those chemicals were newly identified as carcinogenic. Ruthann Rudel, director of the Institute and author of the study, concluded, "The way that chemicals are tested now, we really are missing breast-related effects. We have to do a much better job of testing for these effects when we test chemicals." Prior to this, in 2014, the Institute conducted a study that showed seventeen types of chemicals that "should be top targets for breast cancer prevention" due to their risks when present in women's breast tissue. They included chemicals found in common substances such as gasoline, vehicle exhaust, flame retardants, stain-resistant textiles, paint removers, and even disinfection products in drinking water.

If you are reading this book, chances are you've suffered from at least one of the conditions I've mentioned so far. Maybe as a result of plastic surgery or a mastectomy, your lymph nodes have been compromised. Maybe you've been battling with your weight and insulin levels for some time. Chances are, you've already been exposed to traditional western practices such as lymph massage, water pills, hardcore prescription medications, or what I call "rack-em, stack-em, and crack-em" chiropractic work to alleviate your symptoms. You've tried everything, but nothing seems to deliver permanent relief. You've lost money, time, and energy. Most importantly, you've lost hope. I know what this feels like because a long time ago, I was there too.

My Own Journey Toward Relief

Back in my early twenties, I was cruising through the Midtown Tunnel one night when a car in the next lane swerved into mine, causing me to smash my face into the steering wheel. I developed a severe temporomandibular joint dysfunction (TMJD) problem that didn't go away. TMJD is when a person experiences pain in the joints that act as a hinge for the jaw, connecting it to the rest of the skull and/or the muscles that control them. It's a widespread condition, affecting over 10 million people in the US, and one of the worst parts is that doctors often don't know the exact cause, attributing the pain to some combination of genetics, arthritis, injury, and teeth-grinding.

For four years, I walked around with jaw and neck pain and chronic headaches, popping Motrin pills like they were M&M's until finally, I had to see a gastroenterologist for an ulcerated stomach. I tried dental reconstructive surgery and a ton of different gizmos and gadgets to fix the problem, but nothing helped. Finally, one day, my girlfriend introduced me to her chiropractor, who had just begun to practice Directional Non-Force Technique Chiropractic. After one adjustment, my world was officially rocked. For the first time in eons, my headache went away. Over the next few years, I received treatment while simultaneously studying the technique myself. I knew that I wanted to give others the hope that this therapy had given back to me. So I became one of only 150 Directional Non-Force Technique Chiropractors in the country, practicing this unique method that guarantees lasting results in an average of six sessions.

DR. LORETTA T. FRIEDMAN

Getting to the Heart of the Matter

The *Lymph-Link* is more than just the title of this book. It is the answer to almost all systemic inflammatory issues, whether they are underlying or creating a great amount of discomfort. The state of your lymph is your body's best indication of overall health. Very few institutions promote breast health in this country, which is at the heart of lymph health for women. I'm here to tell you there are effective ways to take care of your breasts and general lymphatic health on an ongoing basis that most doctors might not even know about. Take, for example, the yearly-prescribed mammogram. This Holy Grail of cancer detection can actually miss a lot. But a good thermography, which shows the neogenesis of new blood vessels in the body, can be pivotal in detecting cancer. The FDA even recommends it in addition to mammography for detecting breast cancer. And yet many people have never even heard of it!

Many doctors all over the country are misdiagnosing women left and right. Stacey was a twenty-year-old college student who drove all the way from Ohio to Manhattan to see me for strange swelling in her neck, torso, and legs. Believe it or not, her hometown physician had told her she was probably just eating too much. To add insult to injury, he recommended that she also take a pregnancy test! When this poor young woman came into my office, I diagnosed her with systemic generalized lymphedema and started her on the Lymph-Biologics™ protocol right away.

Lymphedema is thought to affect as many as 250 million people around the world, and it can be caused by any sort of disruption to the functioning of the body's lymphatic system. Yet it didn't occur to this poor woman's doctor even to consider that it might have been what was behind her symptoms. After a week of regular sessions, the swelling in her body went down enough for her to drive home pain-free and with an earful for her doctor.

Don't get me wrong. I wasn't born knowing how to diagnose and prescribe the way I do today. It took a lot of trial and error and even more active listening to find the common thread of toxicity in each patient's story. For example, I used to think my patients were retaining inflammation, because their adrenals were compromised by stress, and their cortisol levels were too high. Cortisol, after all, is the body's main stress hormone, created by the adrenal system when we undergo stress. Of course, some of us experience more stress than others, and a system that is constantly producing cortisol can lead to serious health problems.

But as time passed, I came to see that many of my patients actually had low-stress lifestyles and weren't necessarily anxious people. I also realized that the chiropractic adjustments I was making weren't sticking, and the inflammation wasn't going away. This led me to ask what was still feeding the inflammation. By posing a series of diet- and environment-related questions and checking their cell health, I was able to find that the biggest problem was that their lymphatic systems were out of whack. And the biggest culprit in their poor lymph health was toxicity.

Even Oregon Health & Science University cites cancer treatment as the most common cause of secondary lymphedema. That includes chemotherapy, the introduction of chemical treatments into the body to fight cancer. And as we know, when a chemical taken into the system does harm to the body, it is at that point considered a toxin.

Time to Bang the Drum

I have been doing Lymph-Biologics™ on my patients for more than twenty-two years, witnessing some of the best results you'll find for lymph-related conditions. The device I use in my office combines light and noble gases to create an electrostatic field that, when moved with the right technique and rhythm, improves lymphatic flow and accelerates detoxification of the tissues.

In addition to Lymph-Biologics™, I also prescribe a regime of metabolic detoxification that includes diet change, supplements, and the removal of toxins from my patients' environment. Typically, my patients experience relief almost instantly, and, more importantly, their results last a long time, if not their entire lives.

For years, whenever I talked to my patients about their toxicity levels, most of them looked at me as if I had six heads. I was telling them something that none of their physicians had even hinted at that their cancers, autoimmune diseases, and the general accumulation of lymph fluid in their bodies was caused by a crazy amount of toxins in their lymphatic systems.

After the 2014 Silent Spring Institute study confirmed what I already knew about the link between toxicity and breast disease by identifying household chemicals that could be risk factors for breast cancer, I started banging my drum. Very. Loudly. I reached out to places such as the Lymphatic Education Research Network, sending them information about Lymph-Biologics™ and its success in treating lymph-related issues. I did the same thing with all of the hospitals that were offering alternative programs in the metropolitan area, institutes such as Beth Israel Deaconess Medical Center, Weill Cornell Medicine, Memorial Sloan Kettering, and NewYork-Presbyterian Hospital.

I even contacted all of the top functional medicine doctors around. No one would give me the time of day except the late Dr. Mitchell Gaynor and Dr. Sheldon Feldman. Dr. Gaynor took my phone call immediately and was interested in my lymph work and impressive results. At the time, I wasn't aware of how high on the totem pole in the oncology world he was or that he was the one who introduced turmeric, B12, and Co-Q10 to mitigate symptoms and enhance chemotherapy. Unfortunately, Dr. Gaynor passed away before we could establish a working partnership. Dr. Feldman and I met after I saw a patient of his, who was in excruciating pain after months battling breast cancer. After one visit, her pain went from a 20 to 2 on a scale of 1 to 10. The only question this poor woman asked was *Why*? *Why didn't anyone tell her about this treatment until now*? I told her the reason was that nobody knew about it. I also encouraged her with the fact that she now had the opportunity to share with other pain sufferers what she'd discovered about Lymph-Biologics™ and to help educate her doctors about it.

At the same time, I was seeing more and more patients displaying acute signs of toxicity that coincided with their breast and lymph diseases. I listened to their stories, I witnessed their pain, and, most importantly, I asked the right questions to get to the bottom of their health issues.

The truth is, I'm good at what I do. But I'm only as good as the tools at my disposal. I've cracked some of the toughest medical cases around. What I do is not rocket science, but it does take test results and time to approach each patient holistically. Unfortunately, many patients are hindered by limited health insurance plans. High deductibles and low flexible spending accounts are their biggest opposition to the basic human right to good healthcare.

I decided a long time ago to bypass the middleman and go straight to the folks who needed to hear what I had to say. I started posting blogs and producing YouTube videos, educating my viewers and challenging them to take their health into their own hands.

Turning Over the Rocks

Then, the pandemic struck. If there's anything that COVID-19 has taught us, it's that toxicity is a real thing. It was one issue when elderly and obese people and other "at risk" patients developed lethal cases, but when the younger generation started to get hit hard, I knew it was connected to their toxin levels. People in their twenties, thirties, and forties, who should have been able to weather the COVID-19 storm the best, got much sicker than anyone had expected. In many cases, they died. Even athletes who were considered in optimum health but who had been drinking mercury-treated sports drinks since they were kids became vulnerable to death.

In fact, *Food and Chemical Toxicology* even published a paper arguing that "immune system degradation from multiple toxic stressors (chemical, physical, biological, psychosocial stressors) means that attribution of serious consequences from COVID-19 should be made to the virus-toxic stressors nexus" and that "for long-term pandemic prevention, toxicology-based approaches should be given higher priority than virology-based approaches." The seven international authors of the article came to the same conclusion that I had, namely, that "The most severe consequences from COVID-19 and influenza stem from a degraded/dysfunctional immune system, and the exploitation of the degraded immune system by the virus." And what is the cause of that degraded/dysfunctional immune system? I am sure that by now you have some idea. The authors blamed a combination of many toxic stressors, including lifestyle choices, treatments for other illnesses, biotoxins, social factors, and a long list of environmental factors including chemicals found all around us, such as "microplastics, heavy metals, pesticides, nanoparticle. . .fine particulate matter, [and] air pollution."

The question isn't "Am I toxic?" It's "How toxic am I?" This book will help you find the answer. As you read it, you will explore your own breast and lymph health and find better ways to detoxify and prevent disease. You will also discover alternative treatment methods for lymph-related ailments. Most of all, you will learn how taking charge of your own health is the only way to find lasting solutions.

Everyone has the right to know how their bodies should be functioning and the daily assault that the toxins in our environment and diet are carrying out on our bodies. Traditional medicine doesn't turn over all the rocks. It's often based on vague systems such as blood tests that miss the heart of the matter and ignore the benefits of more nuanced tests for elements such as environmental sensitivities, single nucleotide polymorphisms (SNPs, the most common type of genetic variation, which occurs throughout all of our DNA), cell age, and more.

Finally, I believe it's up to every person to find their own unique path toward health. My hope is that with this step-by-step guide and reference book, I will help you to do just that.

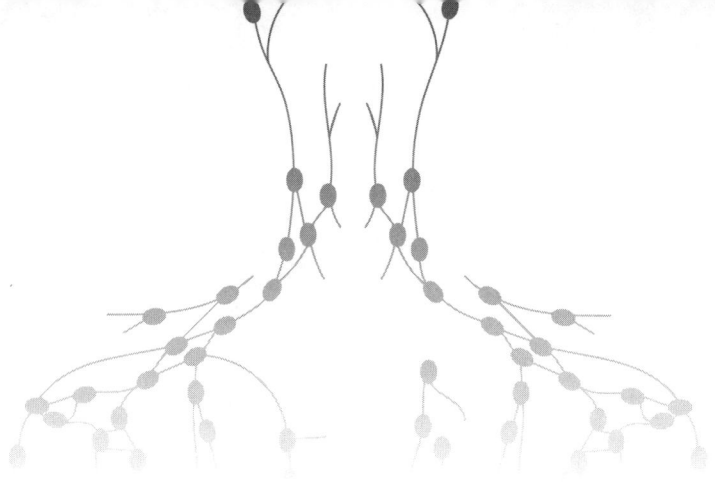

CHAPTER 1

Lymph—The Dirty Little Secret Your Doctor Will Never Talk About

More than 10 million Americans have been diagnosed with Lymphedema—that's more than the number of people with muscular dystrophy (Jerry's Kids), ALS (Lou Gehrig's disease), multiple sclerosis, Parkinson's, and AIDS combined—and I believe that tens of millions more suffer from the condition every day. And that's in this country alone! When the pandemic of 2020 hit, and people all over the world started calling me for telehealth appointments, I saw firsthand what I'd always suspected: another global health crisis has been among us for decades, plaguing our immune, digestive, and circulatory systems.

The pandemic of toxicity is even worse because no one can see it, so no one is talking about it. As my patients from Saudi Arabia, Scotland, Brazil, and many other countries have confirmed, people all over the world are fighting invisible toxins from our external environment and the foods we eat, the water we drink, and the air we breathe.

What does the lymphatic system have to do with our bodies' toxicity levels and many common ailments? Everything. What are most medical providers doing about it? Very little, if anything. As you'll soon learn from this book, if you have lymphedema or lymphoma, then you have toxicity. The problem is, until a person's immune system is challenged, toxicity can go virtually undetected. I am convinced that the COVID-19 long haulers are experiencing prolonged symptoms of the virus, even to this day, because of their abnormal toxicity levels. Foggy brain, insomnia, redness and swollenness in the tips of their fingers and toes, skin issues, a lack of appetite, and a loss of the sense of smell and taste are only some of the symptoms that indicate high levels of toxicity.

This is not a risk that we can dismiss lightly. As we have all learned from the news media over the past months, even being vaccinated against COVID-19 is not an absolute guarantee against developing the disease. And 25 to 33 percent of those who do come down with it develop long-haul COVID-19 symptoms in some form or other. What's more, that statistic stayed consistent whether the patent in question developed a serious case requiring hospitalization or the lightest barely-detectable infection. Though fatigue and shortness of breath are the most common long-haul symptoms, researchers seem to agree that everyone reacts differently to the illness over the long term.

And that's the optimistic view. One study showed that 24.9 percent of COVID-19 patients had at least one symptom a year after their symptoms began. Amid all the hullabaloo, media coverage often seems to overlook the fact that it's a virus that disproportionately affects women. Women, for example, are 1.4 times more likely to report fatigue or muscle weakness from it. We are three times as likely to have lung problems after twelve months, and twice as likely to report experiencing anxiety and depression.

Clearly, COVID-19 is tightly bound up with the presence of toxic substances in the body with the breasts and lymphatic system.

In the next chapter, I'll talk all about common toxins in our food, water, and the air we breathe. For now, what's important to understand is that lymphedema and lymphoma are impossible to talk about without also discussing the issue of toxicity. But before we get into all of that, let me first introduce you to the lymphatic system itself.

Lymph—What is It?

Simply put, the lymphatic system is the body's superhighway disposal system for trash. Once your white blood cells work to eradicate infection and disease, the resulting debris travels through tissues and muscles via lymph fluid contained in an intricate network of vessels. All throughout this network are bean-shaped structures called lymph nodes, where this debris is dumped. Altogether, your body has about 160 lymph nodes in the head and neck region, 300 in the trunk of the body, and hundreds more in the lower extremities.

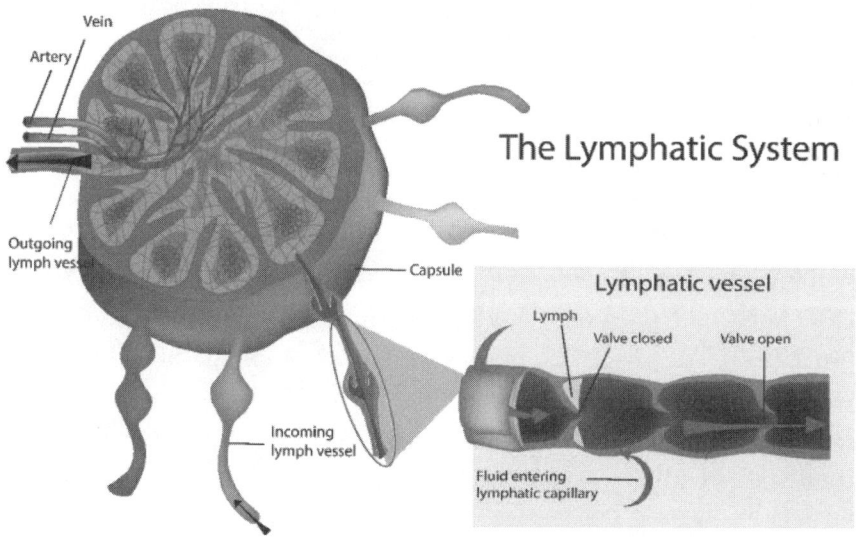

Sakurra / Shutterstock

Depending on what type of waste is being deposited, your lymph nodes can become inflamed or irritated and react by swelling up and becoming hard. Alternatively, they can become extra tender and highly sensitive to touch. This type of reaction often happens in the breast tissue, which contains dozens of lymph nodes. Contrary to popular belief, lumpy, bumpy, and tender breasts are not a normal part of a woman's menstrual cycle but a clear indicator of toxins in your lymphatic system (more on this in Chapter 9).

There is nothing mysterious or sexy about the lymphatic system. It's simple: next to every artery and every vein is a lymphatic vessel. You'll never read an article about it when you're waiting in line at the grocery store. Nor will your doctor ever bring it up—unless there's a problem. But don't mistake simple for insignificant. As you'll soon learn, the lymphatic system is one of your body's most powerful ways of clearing out toxins and maintaining good health. Yet, as I said, unless uncomfortable symptoms start to occur, most people don't even know it exists. In many cases, that's because they have been told by trusted sources that they should ignore what I consider to be a clear sign of lymph dysfunction. Even mainstream medical institutions such as The Women's Clinic of Texas have acknowledged the role of environmental toxins in causing breast lumps, yet they continue not to encourage sufferers to seek remedies and eliminate these toxins from their systems.

Listening to Your Lymphatic System

Not too long ago, Darlene, a software specialist in her forties, came to see me, because the right side of her jawbone was inexplicably swollen. She had already visited several medical practitioners, including a TMJ specialist, who, oddly enough, was the same one who treated me thirty-five years ago after that devastating car accident I told you about in the introduction. This specialist had given Darlene all kinds of bite plates and devices to alleviate her pain and swelling, but nothing worked. When I applied Lymph-Biologics™ to her (more on this three-part, oscillating light treatment in Chapter 7), I noticed that the lymphatic fluid in her face was moving around, but not draining.

I've seen cases like this countless times: patients arrive at my office desperate, having gone through several rounds of unnecessary medical and dental treatments for their misdiagnosed ailments. Sadly, this type of rigmarole is often unavoidable since most doctors are trained in only one area and, therefore, can only see a patient's symptoms through that specific lens. *You got pain in your jaw area? Then, it must be caused by either A, B, or C,* says Dr. So-and-So. That A, B, or C varies greatly, depending on which specialist you visit. Doctors may be looking under the hood, but they're only looking at certain parts. More importantly, most of them aren't paying attention to the quality of the fluids that are running through the pipes and valves—the stuff you can't see with the naked eye.

Poor Darlene's experience was no exception. She was exhausted and feeling hopeless. We did a cell test in my office, as I do with every patient, which gave us immediate answers about what was really happening in her body on a cellular level. A cell test is also known as Aging Analysis. It works on either what we call the basal functions of the cell—those that all cells share—or on special functions that only certain cells carry out. I look for effects in the function of the cell caused by certain substances. And as we all learned in school, if there is an effect, there has to be a cause. If the cells are reacting as we know they do when affected by certain toxic substances, then we know that those toxic substances are present in the body.

As I expected, Darlene's root problem had nothing to do with structural misalignment but rather with the fact that her toxicity levels were off the charts. If I'd told her six months earlier what I told her at her visit—that the swelling in her jaw was caused by a malfunction of her liver—she would have looked at me like I had two heads. But that day in my office, she was ready to hear anything that could provide some relief.

I explained to Darlene that what was going on in her body was *enterohepatic recirculation* (EHRC), a fancy way of saying that her waste removal system was off. As I defined this term and started to teach her about the important waste removal processes that a healthy body performs on an ongoing basis, I watched her demeanor start to change and a glimmer of hope start to shine in her eyes. This moment of understanding, when a patient begins to comprehend the inner mechanisms of their own body, is a constant reminder of why I do the work I do. Most of the time, patients come to me with almost no knowledge of how their bodies are supposed to function. Most of my job, therefore, is to educate them. As the saying goes, knowledge is power. In my practice, more specifically, knowledge is the power to heal yourself.

To understand how EHRC works, you need to know that the body has primarily two phases of waste removal: Phase One, when the liver releases through bile all of the toxins our bodies absorb, and Phase Two, when those toxins get removed through the urine and feces. Without getting too technical—for that, you can read a medical journal—EHRC happens when Phase Two isn't working at full capacity. In Darlene's case, as waste kept recycling back to her liver, Phase One needed to overcompensate by working even harder to break down the toxins. In essence, her liver was dumping more and more toxic debris (mostly heavy metals) into her system, while almost none of it was getting released out of her body. The result, no pun intended—and pardon my French—was a real sh*t show.

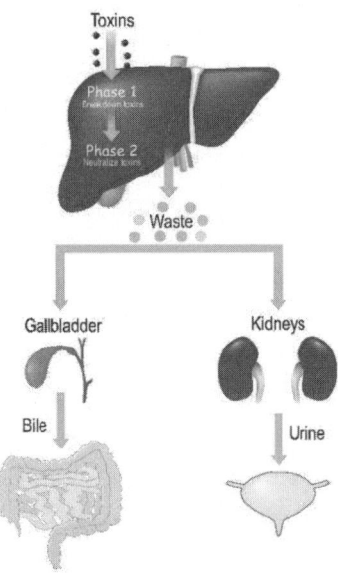

© **Designua | Dreamstime.com**

The heavy metals in her system were inflaming her lymph nodes, creating the uncomfortable swelling in her jaw. I first needed to make some adjustments to her neck, which had gotten completely out of alignment, due to the pain her body was shifting to try to alleviate. Realignment was important to help the lymph fluid in her jaw start to flow properly again. But the real culprit was the level of heavy metals in her system. For that, she needed a regimen of heavy-duty antioxidants, such as liposomal glutathione, and a clean, organic diet.

After discussing her daily life, I learned that Darlene's stress level was high too, which wasn't helping. In addition to her demanding job, she exercised regularly and looked to be in good shape. With her productivity and physical fitness, no one would ever have guessed that she was so toxic and that her Phase One and Phase Two liver functions were in the toilet. I started her on a regimen of ashwagandha and rhodiola with ginseng to support her adrenal glands. I also advised her to spend fifteen minutes a day sitting in meditation before bedtime. Lastly, I talked to her about getting off the subway two stops earlier each day so she could get twenty minutes of cardio in on a regular basis.

Taking a Holistic Approach to Your Lymphatic Health

As you can see from this one example, getting to the bottom of a patient's health issues requires some detective work that most doctors do not have the time or resources for. Quite frankly, some doctors who may have the time and resources just don't care enough to track the clues. Instead, they choose from a handful of diagnoses and provide treatments that are (usually) benign enough not to create more damage. These treatments may or may not alleviate some symptoms, and they never deal with the source of trouble.

Even in the mainstream world that doesn't jive with holistic approaches to health care, it's universally recognized that misdiagnosis is a serious and urgent problem in the health care world. By some estimates, misdiagnosis in hospitals alone kills 80,000 people per year. Just imagine what that number would look like if we added in deaths that were the result of misdiagnosis in the offices of primary care physicians or specialists—many of whom the patients should never have been referred to in the first place!

DR. LORETTA T. FRIEDMAN

Patients like you who suffer from lymph issues are especially vulnerable to incidences of misdiagnosis. For example, scientists have recently identified a disease called indolent T-cell lymphoproliferative disease (T-LPD) of the gastrointestinal tract. That's a mouthful, but, in simple terms, this disease causes lesions of the gastrointestinal tract and symptoms that, even to the trained eye, closely resemble those of non-Hodgkin lymphoma. As a result, doctors often begin treating the patient for cancer. Cancer treatments kill healthy cells in their effort to kill cancerous cells, which means that the misdiagnosis leads to innocent patients' healthy tissue being destroyed in an effort to fight a cancer that does not exist. What's more, indolent T-LPD does not respond to cancer treatment—which means that those who have it will continue to suffer from their symptoms without relief if they are mistakenly treated for non-Hodgkin lymphoma instead.

What does this all add up to? Misdiagnosis is a silent plague. And it's one that virtually nothing is being done about because most doctors don't even consider it to be a threat. After all, how can you realize the danger of something that you don't even think exists? That's right: while cancer misdiagnosis rates as a whole are at about 28 percent, a survey of doctors found that a 60.5 percent majority of them would estimate that misdiagnosis rate as somewhere between zero and ten percent. Of course, nobody likes to admit that they themselves are making mistakes. But doctors need to do it when the stakes involve human lives. And here's the kicker: the *BMJ Quality & Safety* medical journal has found that the most misdiagnosed cancer is—you guessed it—lymphoma. If that much misdiagnosis is going on where cancer is concerned, just imagine that amount that is happening involving non-cancerous lymph diseases, which doctors tend to treat less seriously than they do cancers.

Helping a patient to take the reins of their health into their own hands is a holistic approach that medical school never talks about, but it is what my own practice is all about. Considering how many lymph-related medical problems get misdiagnosed each year, I believe that teaching people about the Lymph Link is a matter of life and death. It's why I wrote this book. Think of all the time, energy, and money people would save trying to find out what is wrong with them if their doctors did a simple cell test or urine analysis on their first visit to determine their toxicity levels and cellular functionality! Clearly, when it comes to medical detective work, we are missing the boat. But why?

We live in a fast-paced, consumer society that values quick, short-term results over long-term solutions and lifestyle changes that will truly make a difference. Consider all of the over-the-counter medications that are at our disposal at every pharmacy, deli, and truck stop. And that's not saying anything about the countless prescribed medications that keep pharmaceutical companies raking in millions of dollars every day. Conversely, learning to slow down, listen to clues your body is showing you, and learning new habits that promote health are not activities that stimulate the economy. They also hardly fit into our jam-packed schedules that allow very little room for relaxation. In western culture, self-reflection and self-care are considered luxuries.

Yet the rates for cancer and heart disease, conditions both caused by stress, are off the charts. Statistics predict that by 2035, 45 percent of people in the United States will suffer from some issue related to heart disease and that the total number of cancer cases in the country will rise 49 percent from 2015 to 2050. And that's not just due to the aging of the population. A recent study showed that between 1973 and 2015, people between the ages of fifteen and thirty-nine showed a shocking 30 percent increase in cancer rate. Human biology has not changed in the past decade. Is it any wonder, given this information, that I look to the factors of increasing stress levels and of environmental toxins for an explanation? These are not warning signs we can ignore.

Very often, lymphedema is an early warning sign that we are heading toward some of these more lethal conditions. It is the body's way of signaling that it is overloaded with stress and toxins—and is a clear cry for help. Most doctors are trained in practices that are aimed at quieting the noise so that patients can move on with their fast-paced lives, despite the deadly implications. I, on the other hand, teach my patients to slow down and listen to it.

DR. LORETTA T. FRIEDMAN

Sharaf Maksumov / Shutterstock

Lymphedema: The Disease With a Thousand Faces

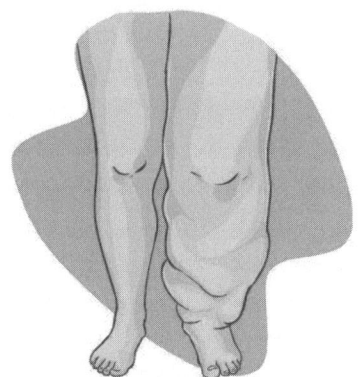

Drp8 / Shutterstock

If you look up the definition of lymphedema online, you'll find something like this: *swelling in an arm or leg caused by a lymphatic system blockage.* While this description is accurate, it doesn't tell the whole story. That's because lymphedema is a complex human disease that has been largely neglected by health care providers and government health care agencies. Only in the last decade has more effort been put into gaining insight into the disease. Yet, although progress has been made in fields such as lymphatic imaging and research into the connection between lymphedema, obesity, and other comorbidities, western medicine still has a long way to go in understanding and teaching patients about lymphatic health.

Adding to this lack of education in lymphatic health is the knee-jerk reaction doctors tend to have to grab the quickest fix to a problem, which is often a knife. Over the last decade, there has been a dramatic upsurge in surgical interventions for lymphedema patients. These include procedures designed to reduce the likelihood of lymphedema, debulking operations, and microsurgical interventions to correct impaired lymphatic vascular function. Still, while these procedures sound promising to patients who are desperate for relief, very little evidence exists that they are effective in the long term. Moreover, very little is understood by doctors about much less invasive yet highly more effective modalities that almost never are offered to their patients as alternative options.

There's an old saying: to a hammer, everything looks like a nail. Well, to a surgeon, everything looks like it needs surgery—including lymphedema. But any surgery carries risk with it, and taking that risk ought to be justified. Lymphedema surgery is no exception. But rarely are its disadvantages discussed. Search the internet for data, and you're likely to find a lot of very positive information about surgery for any condition—but it's worth considering that most of the sites you find are written by surgeons, who are likely to think that surgery is the answer to everything and have a vested interest in more people opting to get these elective surgeries.

Just the other day, Terrence, a fifty-seven-year-old plumber from Georgia, scheduled a video call to show me his left foot and lower leg that had been swollen for three months. He told me that although his doctor already had scheduled him for surgery three days later, his wife had asked him to reach out to me to get a second opinion. When I asked Terrence what exactly the doctor was planning on doing to his foot and lower leg once he was under anesthesia, he had no clue how to answer me. X-rays had already revealed that there was no detected blockage in his vascular system, nor was there a localized pocket of fluid to be released. As with almost all lymphedema cases, the swelling in Terrence's foot and lower leg was due to an insidious onset of lymphedema caused by a buildup of toxins. After briefly discussing his diet with him and discovering that he also suffered from many digestive issues, I suspected that the main culprit for his lymphedema was Leaky Gut Syndrome.

That's a condition that is essentially just what it sounds like—some medical texts refer to it by the longer name of "increased intestinal permeability"—and it leads to symptoms including aches and pains, food sensitivities, gas, cramps, and bloating (not to mention, of course, lymphedema, as in the case of Terrence). While mainstream medical practitioners have generally thrown up their arms in the face of leaky gut syndrome, professing that they know neither what causes it nor how to treat it, they do make one central admission: the cases that they can identify are those linked to "certain types of drugs, radiation, or food allergies." Of course, drugs, radiation, and/or foods that a person is allergic to would constitute foreign substances to the body which cause an adverse reaction when they are introduced—or, in other words, toxins!

I recommended to Terrence's wife that under no circumstance should he have surgery on his leg. What was the doctor going to do? Perform an exploratory on his calf? As I explained, there was nothing there to operate on. "If he wants to cause irreversible harm to his leg," I told her, "then he should go ahead and have the surgery. Then, he'll have real problems." I couldn't help but add, "Find another doctor." In my next conversation with Terrence, I told him that in order to treat him properly, I first would need him to fill out a thirty-page patient intake form that thoroughly investigated his diet, environment, and overall health. I would then need to perform some simple tests to discover the main toxins in his body. He would need to come into my office on a regular basis for Lymph-Biologics™ treatment as well to move the lymphatic blockages that were causing the swelling. Finally, he would need to change his diet and also take daily nutritional supplements to help with his metabolic detoxification. Of course, this protocol would require both time and money, two things that a surgery covered by his insurance plan did not require.

Like most western medical procedures, the problem with surgical approaches to lymphedema is that they do not address the underlying causes of a compromised lymphatic system. Most likely, your own doctor has hardly asked you questions about your diet, exercise, stress, and toxicity levels. Nor have they offered you a simple cell test to see how your body is functioning on a cellular level. Instead, they may have handed you a compression garment, prescribed water pills, told you about lymphatic massage, or, as in Terrence's case, recommended surgery. And yet no one in the operating room is asking the pressing question, *How did we get here in the first place?*

What causes lymphedema? Many people develop lymphedema after a surgical removal of lymph nodes. Sometimes, this removal is intentional (say, because of cancer) and sometimes, it's not (say, because of a careless plastic surgeon). However, most of my patients who suffer from swelling and inflammation have never undergone surgery. In nearly all of these cases, the main culprit wreaking havoc on their lymphatic systems is toxicity.

So why hasn't your doctor mentioned anything about your lymph or your toxicity levels? As I said earlier, most doctors are looking at their patient's symptoms through a tiny peephole. I see so many misdiagnosed patients in my office every day because of this limited medical perspective and knowledge. Here are only some of many misdiagnoses for lymphedema that are out there, ranging from the slightly unlikely to the downright absurd: arthritis, Lyme disease, a root canal

> **Signs and Symptoms that Indicate You May be Having Problems with Your Lymphatic System**
>
> - headaches
> - chronic fatigue
> - excess weight and/or water retention
> - skin conditions
> - sinus infections
> - digestive disorders
> - asthma
> - arthritis
> - cancers
> - unexplained injuries
> - lumpy and tender breast tissue
> - lumps under the arms or in the groin region

infection, allergies, unexplained injuries, weight gain, and, my favorite, pregnancy. (Misdiagnoses referenced in the introduction of this book.)

One of the biggest reasons why lymphedema is so difficult to diagnose is that it shows up differently in each person. It's what I call "the disease with a thousand faces." One patient may experience only localized swelling (say, a boil at the tip of their nose) while another patient may develop swelling all over their body (systemic lymphedema) that makes them look like a Macy's Thanksgiving Day Parade balloon. Some patients suffer from extreme pain, and others have no pain at all. For some, the inflammation is ongoing; for others, it's only occasional. The point is, no two lymphedema cases are alike, making the condition a challenging one for doctors to pin down even if they wanted to.

Now that you know more about the lymphatic system and its links to many health issues, it's time to step into the driver's seat of your healing process. In the next chapter, I will discuss the importance of taking charge and asking the questions you need to ask about toxicity and what's really happening inside your body.

What You Need to Know

- Until a person's immune system is challenged, toxicity can go completely undetected.
- COVID-19 long haulers most likely have high toxicity levels that are keeping them from proper recovery.
- The lymphatic system is a vital part of your body's immune system.
- The lymphatic system is an intricate network of vessels through which debris travels via lymph fluid.
- Lymph nodes are bean-shaped structures where this debris gets dumped.
- Your body contains 160 lymph nodes in the head and neck region, 300 in the trunk region, and hundreds more in the lower extremities

- Toxins can inflame the lymph nodes, causing them to swell, harden, or become extremely tender.
- Lumpy, tender breasts are a sign of toxicity
- Since most doctors have training only in a specific area, they have limited ability to diagnose what's really happening in your lymphatic system.
- Natural remedies such as turmeric, ashwagandha, and Ozone therapy can be highly effective in treating inflammation, compromised adrenal glands, and lymphoma. These are almost never prescribed by western doctors.
- Lymphedema is often the first sign that your body contains high toxicity levels.

DR. LORETTA T. FRIEDMAN

arloo / Shutterstock

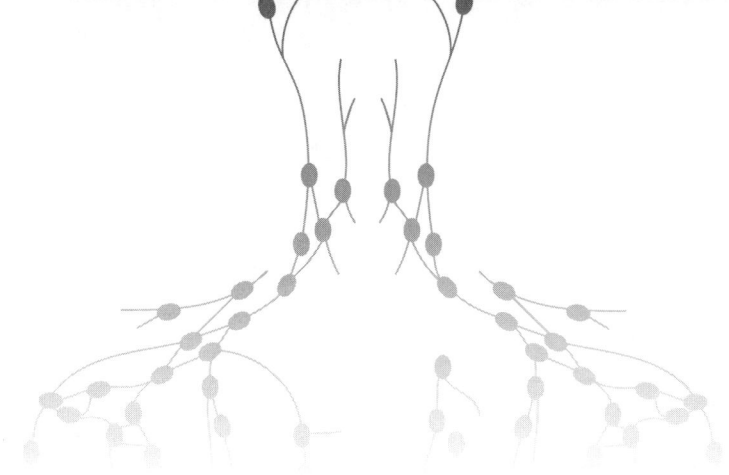

CHAPTER 2

How the Lymph Link Will Help You Find Relief

Stan, a guy in his late seventies, came to my office to be treated for back and neck pain that he'd been experiencing for years. I couldn't help but notice the extremely dark, raccoon-like circles under his eyes—an indicator of a malfunctioning liver—and had to ask if he was a drinker. He told me he wasn't. It was a natural question for me to ask. The liver's main function in the human body is to detoxify substances. It's our first line of natural defense against toxicity. But the liver, like all of our body parts, is susceptible to breakdown if overtaxed. Though we tend not to think of it that way, alcohol is medically considered a toxin, which is why people who drink too much often tend to have problems with their livers. But when Stan told me that he didn't drink, I knew that some other toxin or combination of toxins was at work, causing the telltale sign on his face of liver trouble.

When I asked him what else was giving him trouble, he told me that for thirty years, his doctors had diagnosed him with arthritis in almost all of his joints, and that was the cause of all his pain. I told him that arthritis all over the body was highly unlikely and that the pain and swelling was probably systemic inflammation. His cell test indicated a whopping toxicity level of 38 percent, a number I normally only see in cancer patients.

The test also showed that his cell membranes were shrinking like a raisin due to low levels of Omega-3.

When I suggested a strict anti-inflammatory diet, Stan's wife, who was accompanying him, laughed and said I was barking up the wrong tree, that he'd never comply. He surprised both of us when he followed every word I told him—to the exact T. In addition to his diet, I'd prescribed a high dose of turmeric, a natural anti-inflammatory, that he took three times a day. After a handful of treatments of Lymph Biologics™, not only did Stan's back and neck pain subside, but the dark circles under his eyes got lighter. And the misdiagnosed arthritis? The pain had completely disappeared.

The lesson here is this: if some doctor(s) gives you a diagnosis that feels like a thirty-year sentence of a disease that you can do very little about and find almost no relief for, then question the living daylights out of it. Get a second, third, or fourth opinion. Ask yourself if the diagnosis even makes sense. Then, go to a little tool called the Internet and start to look into information yourself. Go back to your doctor with questions. If they stare at you with a blank face or look annoyed, find a different doctor, preferably one who will conduct a cell test and knows how to interpret it. Be your own detective for your health until things make sense to you.

In Chapter 5, I will discuss the major sources of toxins in our environment and food and teach you some of the ways you can detoxify your body. For now, here's a short checklist of questions I ask all of my patients to help assess their overall health and where toxicity might be finding its way into their daily lives. Use it to evaluate yourself right now.

Toxicity Mini-Quiz

1. Do you take prescription medications?
2. Are you exposed to dust, overstuffed furniture, tobacco smoke, mothballs, incense, or varnish in your home or office?
3. Have you been diagnosed with thyroid problems?
4. Do you eat takeout most of the time?
5. Do you consume diet beverages?

6. Are you sluggish or tired?
7. Are you exposed to nail polish, perfumes, hair spray, or other cosmetics?
8. Do you experience dizziness and headaches?
9. Do you experience hives, rashes, or dry skin?
10. Do you consume conventionally-grown fruits and vegetables that contain pesticides?

Colored Lights / Shutterstock

If you answered yes to three or more of these questions, chances are you have a high level of toxicity. Later in the book, I'll suggest simple ways you can lower that level immediately. What's important for now is to recognize the Lymph Link and begin to connect your uncomfortable symptoms with the number of toxins you are exposed to each day. As is the case with most health issues, the first step to healing is knowing.

Treating Lymphoma—the Better Way

Lymphoma, cancer of the lymphatic system, is a highly-treatable disease. While many patients find positive results with traditional radiation and chemotherapy, some choose an alternative path that can be just as, if not more, effective. Essiac tea is a potent, natural, cancer-reducing drink that many alternative doctors faithfully prescribe to their patients. It was introduced to the world in 1922 by a nurse named Rene Caisse. And no, it's not made from a plant called Essiac. In fact, it's a preparation of several ingredients—burdock root, Indian rhubarb root, sheep sorrel, and slippery elm. Sometimes, watercress, blessed thistle, red clover, or kelp are added. Caisse got the recipe from an Ontario Ojibwa medicine man who had been a patient of hers. And even despite its alternative origins, several scientific studies have been conducted on Essiac tea over the years. One paper, published in the *Journal of Ethnopharmacology*, reported that its data "indicate that Essiac tea possesses potent antioxidant and DNA-protective activity, properties that are common to natural anti-cancer agents. This study may help to explain the mechanisms behind the reported anti-cancer effects of Essiac." The authors of the paper included Stephen Leonard from the CDC, as well as professionals from Montana State University, the UPMC Children's Hospital of Pittsburgh, and the Cincinnati Children's Hospital Medical Center. When even the highly respected medical establishment starts taking a holistic treatment seriously, it's time to take a close look at it. The last line of their research speaks for itself: "Together, these data indicate that Essiac possesses a spectrum of antioxidant and DNA-protective properties common to anti-cancer agents."

Another non-traditional treatment for lymphoma is Ozone therapy. During this procedure, a pint of blood is withdrawn from the patient, infused with O3, then transfused back into their bloodstream so that the Ozone can devour any harmful cells in its path, like a greedy little Pac-Man. This treatment is also highly effective in treating Lyme disease, hepatitis, and HIV. Ozone therapy has been the subject of much alarm in the mainstream, traditional medicine community over the years. They claim that the practice is useless and can even lead to death. The influence of this scientific contingent has even gone so far as to convince the United States Food and Drug Administration into banning the use of Ozone for medical purposes in 2003.

But as any scientist will tell you, Ozone is absolutely and completely natural and is made up of a pure trinity of molecules of oxygen—the same elemental gas that we all need to breathe in order to stay alive. Despite its natural composition, the discovery of Ozone dates back only to the mid-nineteenth century. But already in 1856, doctors had discovered a medical use for it, finding that it was effective in helping them to sterilize surgical instruments and disinfect operating rooms to prevent the spread of disease. In the years since, experts have gone on to study it further, seeking more beneficial effects.

The Journal of Natural Science, Biology and Medicine published a paper on the effects of Ozone therapy by A. M. Elvis and J. S. Ekta, two researchers at the Vivekanand Education Society's College of Pharmacy in Mumbai, India. What Elvis and Ekta found is compelling. They concluded that "Although O3 has dangerous effects, yet researchers believe it has many therapeutic effects… Its effects are proven, consistent, safe and with minimal and preventable side effects. Medical O3 is used to disinfect and treat disease." They explained that the way that the gas works in medical context is by causing bacteria, viruses, fungi, yeast, and protozoa to be deactivated as well as by activating our own natural immune system and by the stimulation of the metabolism of oxygen. They go on to detail the treatment's effectiveness in treating AIDS, SARS, cancer, arthritis, infected wounds, circulatory disorders, geriatric conditions, macular degeneration, and viruses. If you have a treatment that can truly cure all of these terrible ailments, it's criminal not to pass that therapy on to the patients who so desperately need it.

A good holistic medical provider will help you assess which treatments might be right for you by conducting a short battery of tests. When lymphoma patients come to me, for example, I immediately test them for heavy metals, environmental sensitivities, food sensitivities, leaky gut syndrome, and hepatic malfunction to see if they are detoxifying properly. Questions about their diet are an important part of this intake, as you'll see in this next story.

Vladimir, a highly overweight man in his mid-sixties, came to see me at the request of his wife before he started chemotherapy to treat his lymphoma. As is the case for many lymphoma patients, his spleen was very enlarged, and he experienced severe swelling in the chest and neck. When I asked him what he ate every day, he answered that for breakfast, he had a whole avocado and a few eggs with toast. For both lunch and dinner, he ate a big bowl of vegetable soup. I found this diet choice a little curious.

Then, Vladimir's wife chimed in from her chair in the corner of the room. "Ask him how much olive oil he puts in each bowl." Now, my ears were really perked up. Apparently, this guy had a habit of adding three heaping tablespoons of olive oil into each bowl of soup he devoured. When I added it all up, that came to 1,800 calories of fat out of the 2,800 or so he consumed every day! It was no wonder that his toxicity level was off the charts at 42 percent. Doctors have known for a while that excess fat can cause inflammation of the liver. Even those who usually cast a skeptical eye on the role played by inflammation in the illnesses that afflict the human body can't deny this fact. Steatohepatitis is the name for this liver inflammation—and it is known to cause danger to the organ. In a case like this, dietary fat can itself become a toxin! And that is only one of many reasons that it is important for me to investigate a patient's diet when searching for the hidden causes behind symptoms.

Very often, with just a little probing about diet, the writing is on the wall—or, in Vladimir's case, in the bowl. Now, here's what all my patients know who come to see me: if I'm going to ask the questions, you need to be honest with your answers and also be willing to do something about them. Sadly, once I told Vladimir that he would have to change his diet, from the look on his face, I knew I'd never see him again. When it came to his health, he chose to ignore his bad habits and let the professionals (the chemo technicians) do the work.

Vladimir's refusal to do what was called for was the rare exception among my patients. Most of the people who come to see me have reached the end of their rope and are eager to try any new thing that might help. Almost all of them have toxicity levels they shouldn't. In the next chapter, we'll take a closer look at some alternatives to pain management that you might not have considered.

What You Need to Know

- The liver's main function in the human body is to detoxify substances.
- In order to get to the bottom of your health issues, you must be willing to do some detective work and invest time and money into holistic medical care.
- Don't just take one doctor's word for it. Get several opinions and ask questions.
- Dietary fat can become a toxin and interfere with liver function.
- Alternative treatments such as Ozone therapy and Essiac tea can be highly effective in treating cancer patients.

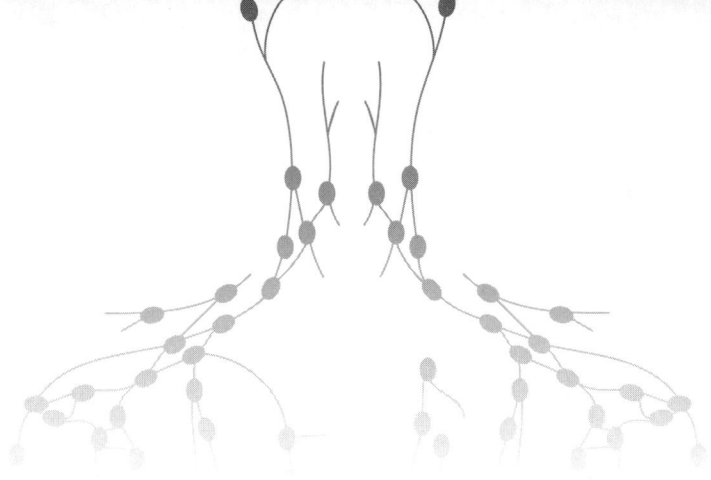

CHAPTER 3

Taking the Driver's Seat to Finding Relief

Considering all of the self-care and treatment that go with it, a diagnosis of lymphedema can feel like a life sentence. But it doesn't have to. It is true that you will need to pay attention to possible complications and secondary symptoms such as skin issues, protein buildup, and an increased risk of infection. In addition, your quality of life will feel compromised with a lower range in motion and visible swelling in your body and face. However, these symptoms do not need to be a permanent part of your life. In fact, I've made it my own life's work to make sure they are not.

The current standard therapeutic intervention is called complex decongestive therapy. It involves manual lymph drainage by a physician and the frequent wearing of compression garments. As you will see in this chapter, this type of treatment is only a temporary source of relief to your symptoms. Doctors often prescribe additional home treatment options such as pneumatic compression pumps that mimic manual massage. But these are bulky, difficult to apply, and require you to sit still for a long time during treatment. Talk about a life sentence!

Lymphedema is a common secondary diagnosis for patients undergoing cancer treatment. Cancers associated with lymphedema, such as breast and prostate, have seen improved survivorship, from 75 percent to 91 percent for breast and from 66 percent to 99 percent for prostate, respectively, from 1975 to 2011. As survivorship improves, long-term side effect risk, particularly for lymphedema also increases with 5–50 percent of cancer survivors developing lymphedema breast cancer-related lymphedema, accounts for 25,000–50,000 diagnoses annually in the United States alone.

Recent news about cancer survival is even better. It can take some time to do full analysis of statistics, but in 2021, an analysis from *CA: A Cancer Journal for Clinicians* showed that, overall, cancer rates dropped by an average of 1.5 percent per year between 2008 and 2017. Of course, every form of cancer is different—and, luckily, it's possible to say that with regard to lymphoma, the news is even better. According to the American Cancer Society, the death rate from Non-Hodgkin Lymphoma has decreased by 2 percent annually over the years between 2009 and 2018—that is .5 percent better than the average for all cancers. That means that we are always learning more about treatments and how to keep people alive for longer. But it does not mean that we can rest on our laurels and ignore those who continue to be diagnosed. And, vitally, it does not mean that we can ignore the often-agonizing symptoms suffered by those who have untreated lymph or breast diseases that fall outside the purview of these studies because they are not considered to be cancers.

The Grim Truth About Pain Management

As I mentioned in the last chapter, conventional treatment for lymphedema is, for the most part, ineffective. Moreover, pain management has been a dismal failure. In fact, it's arguable that the medical establishment has widely ignored the extent to which pain often accompanies lymphedema, seriously reducing the quality of life for sufferers. A 2016 study by three physicians published in *Lymphatic Research and Biology* found that approximately 50 percent of sufferers of lower-limb lymphedema reported experiencing pain. They noted that a previous report had people believing that the number was closer to only 23 percent. It's amazing what we can find if we only look for it! Needless to say, it's impossible to treat someone for pain who is one of the 27 percent of sufferers you don't even know are feeling it.

Many common non-steroidal anti-inflammatory drugs (NSAIDs) such as aspirin, ibuprofen, and naproxen not only offer minimal pain relief, but also pose new, potentially fatal health problems. In the case of aspirin, these include an increased risk of heart attacks, strokes, and gastrointestinal bleeding. Ibuprofen is often seen as the safer alternative, but it turns out that view may be wrong. This common pain reliever has been linked to stomach pain, heartburn, nausea, vomiting, gas, constipation, and diarrhea.

But that's just the beginning of its potential side effects. In other, less-fortunate ibuprofen users, the drug has led to increased risk of heart attack and stroke, decreased kidney function, high blood pressure, ulcers, and bleeding in the stomach or intestine. For a few unlucky people, ibuprofen has even led to liver failure. Using naproxen instead doesn't eliminate your risks, but rather presents a whole new set of them, such as exposure to the possibility of serious similar side effects, including stomach pain, constipation, diarrhea, gas, heartburn, nausea, vomiting, and dizziness. What's more, it is known to have the potential to interact dangerously with a whole host of other medications that a pain sufferer might be taking.

What other options does that leave patients with? Other prescription drugs such as opioids have caused widespread addiction and thousands of cases of overdose. Up to 29 percent of people who are prescribed opioids for chronic pain end up misusing them. That's almost a third—and this number doesn't even take into account the people who start using opioids without ever having been prescribed them. The bottom line is that prescription opioids lead to rampant overdoses, with deaths reaching an all-time high in 2020 of 93,331. That's a huge risk to take, but the lack of resources to treat lymphedema pain leaves patients with almost no other choice. This risky treatment is prescribed by establishment western-medicine doctors all over the country. The real issue here is that in almost all cases, the underlying cause of the pain remains unaddressed.

As I've already said, pain solutions such as arthroscopic surgery are often no better than a placebo or leave the patient in even worse shape, as in the case of failed back surgery syndrome. (Yes, this last issue is so common it even has a name!) Unfortunately, it's just what it sounds like. Unfortunate sufferers go into surgery, hoping that it will cure their often-debilitating aching back, and the result is nothing but a potentially risky procedure, pain from the surgery, and a still-hurting back. Estimates say that failed back surgery syndrome may occur in up to as many as 40 percent of people who go under the knife for back pain. Many of them go back to surgeons in hope of better chances, but are often disappointed since the syndrome becomes more and more common with repeated surgery.

Yet treatments such as back surgery are promoted as the best ones for chronic pain. Over 100 million Americans continue to suffer from mistreated chronic pain each year. And no one until now has come up with a better alternative.

Choices in Pain Management

The fact is, pain has many causes, ranging in nature from musculoskeletal, nutritional, mental, emotional, and spiritual. Several of these types of pain are best addressed by alternative medicine that offers safer, more natural relief than pharmaceuticals. Popular alternative treatments include cannabis, kratom, Omega-3 supplements, low-level laser, and lymphatic drainage. Sadly, most western physicians are either uninformed about these treatments, have conflicts of interest that have to do with insurance and business partnerships with pharmaceutical companies, or are afraid of having their practices challenged by these entities. For these and many other reasons, most doctors will not bother to educate themselves or their patients about these alternatives.

Cannabis is widely known as a recreational drug, but it is gradually being legalized in more and more states—sometimes for recreational use and sometimes only as medical marijuana. That means, of course, that its medical advantages are indisputable even in the established world of western medicine. In fact, the leading journal *U.S. Pharmacist* reported that 85.5 percent of patients who used medical marijuana showed either substantial or conclusive evidence of its having reduced their pain.

Cannabis's pain-relieving effects are real. In fact, a study published by the *Hawaii Journal of Medicine and Public Health* examined to what extent the use of marijuana decreased people's pain. They used the 1–10 pain scale that doctors in hospitals will often employ when evaluating patients. What they found was dramatic: patients' pain decreased an average of five points, from 7.8 to 2.8—a 64 percent relative decrease in average pain. But despite cannabis's effectiveness at relieving pain, many doctors are less likely to prescribe or recommend it because they don't have the financial incentives or relationships with its distributers that they do with pharmaceutical companies.

While just about everyone knows what marijuana is, you aren't alone if you were sitting in front of this book, shaking your head at the unfamiliar word kratom. It may sound to you like some strange and exotic chemical. But in fact, it's as natural as a tree. This tree grows in Southeast Asia, and its leaves are known to be either a stimulant in low doses or a sedative in high doses. It's legal in the United States and can be chewed, made into a tea, or taken as a pill. And, notably for lymph pain sufferers, 91 percent of those who use kratom in the United States do so for the purpose of pain relief. So many people don't take a remedy for pain and continue to do so if it isn't helping them. And fortunately, science backs them up. A 2019 scientific study done at Portugal's Universidade da Beira Interior and published in *Medicines* showed that kratom's two active alkaloids had not only pain-relieving, but also anti-inflammatory properties. I'm sure you can imagine what an effect that might have on a painful case of lymphedema.

You've probably heard about Omega-3. You may know that it's a kind of fatty acid found in fish oil, but it's a strong possibility that the medical establishment has not made a point of educating you on how it can be an effective tool against pain. But even such a mainstream source as the University of Pittsburgh Medical Center points out in its literature that science has recognized that consuming one to three grams of fish oil per day can help reduce joint pain, and that Omega-3 can reduce inflammation—a major culprit in pain the world over, especially in relation to lymphedema.

Low-level laser therapy is a tool for fighting lymphedema of an entirely different character. Unlike Omega-3, kratom, and marijuana, it is not a substance provided for us by Mother Nature, but rather a miracle of modern technology that has provided real relief for many lymphedema sufferers. It's just what it sounds like but not quite so frightening. A lymphedema therapist uses a small, hand-held laser device to apply lasers to the affected skin for short intervals. The patient feels nothing. And according to scientific studies, the therapy benefits the patient by breaking down scar tissue, increasing range of motion, reducing tightness, and reducing the volume of the affected area. Naturally, reduction in these effects can also lead to a reduction in the level of pain. The theory is that it works because laser light increases the flow of lymph, thereby decreasing the amount of protein and tissue in the lymphedema fluid—meaning scar tissue can't adhere as well to the healthy tissue underneath. Because of its demonstrated effectiveness, even the United States Food and Drug Administration has approved low-level laser therapy as a treatment for lymphedema.

Lymphatic drainage is another often-overlooked weapon in our arsenal against lymphedema. It's actually a form of massage—and one that can have wonderful benefits for sufferers of lymphedema pain. Gentle pressure is applied not just to the area affected most by lymphedema but to the entire lymphatic system area. Since it is not a substance but a massage technique, this method of relieving lymphedema symptoms and pain costs nothing except for the fee paid to a skilled practitioner. And yet many doctors fail to recommend it and open this avenue of relief to their patients.

Letting Go of the Story

Still, another factor exists that is keeping Americans from getting the relief that is just within arm's reach. Many chronic pain sufferers experience a phenomenon that I call "getting stuck in your story of pain." Recently, I went to see a one-woman show to benefit the Lymphatic Education & Research Network (LERN). The production was written and performed by a woman in her thirties who had suffered from lymphedema since she was twelve years old. The play chronicled her decades-long story of misdiagnoses, multiple hospitalizations, and various medication protocols, none of which could alleviate her systemic swelling and inflammation.

After learning about her grueling experience, I made a beeline for backstage once the show was over to offer the woman comprehensive treatment in my office at no charge. This would include cell testing, chiropractic work, Lymph-Biologics™, nutritional guidance, and recommendations for supplements. The woman seemed elated at first and gave me her contact information right away. She promised to take me up on my offer as soon as the show was done running. I expected her to call within a few weeks, but, surprisingly, I never heard from her. After several months of trying to reach her, I finally got her on the phone. I explained the protocol I use for my lymphedema patients and assured her that her condition was highly treatable. Apparently, when she realized that it would take several office visits and changes to her lifestyle to get the relief she said she wanted, she told me it was too difficult to get from her house on Long Island into the city for treatment.

Like many people, this young woman had been telling her story of pain and discomfort for so long that she'd grown too comfortable with it to exchange it for new slight inconveniences. After all, she'd even established herself as a performer through her story of suffering. Why would she want to give that up? It's a strange yet common response I get a lot in my practice. The truth is, most people suffering from chronic lymphedema require a certain amount of resilience and optimism that years of poor health and even poorer healthcare can easily zap out of a person. Unfortunately, as we saw in the first chapter with Terrence from Georgia, all too often, it is easier to remain stuck in your story of suffering than it is to rewrite the script.

In the next chapter, I will share how Directional Non-Force Technique taught me to think outside of the box, and why it's so important to do the same regarding your own health care.

What You Need to Know

- As cancer survivorship improves, long-term side effects risk, particularly for lymphedema, also increases, with 5 to 50 percent of cancer survivors developing lymphedema.
- The medical establishment has widely ignored the extent to which pain often accompanies lymphedema.
- Many common non-steroidal anti-inflammatory drugs (NSAIDs) not only offer minimal pain relief, but also pose new, potentially fatal health problems.
- Up to 29 percent of people who are prescribed opioids for chronic pain end up misusing them.
- Over 100 million Americans continue to suffer from mistreated chronic pain each year.
- Most western physicians are either uninformed about effective alternative treatments, have conflicts of interest with these treatments, or are fearful of having their practices challenged by pharmaceutical and insurance companies.
- Popular, effective alternative treatments include cannabis, kratom, Omega-3 supplements, low-level laser, and lymphatic drainage.
- Many chronic pain sufferers get stuck in their stories of pain that keep them from seeking the help they need.

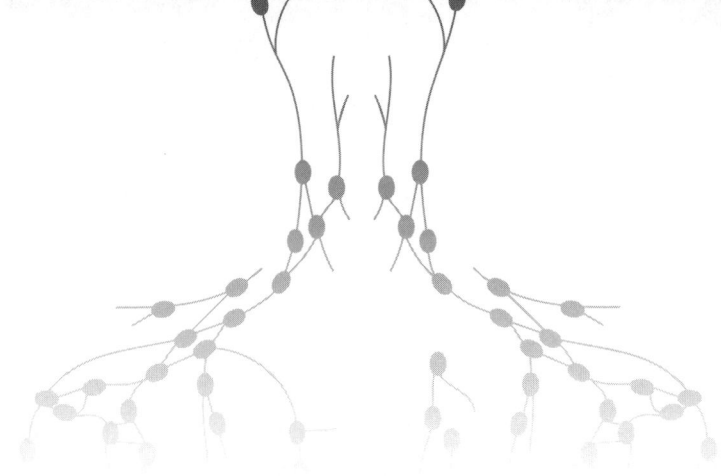

CHAPTER 4

How I Cut through the Bullsh**—and You Can Too

Believe me, I should know. Thirty-five years ago, I was on a mission to find any type of treatment to alleviate the excruciating sciatic pain in my right leg that had developed suddenly while I was working as an orthopedic nurse at the Hospital for Joint Diseases. I also still had major jaw pain from the car accident I mentioned earlier. I'd already tried all kinds of lumbar supports and belts from the orthopedic hospital where I was working at the time, but nothing helped. Then, my friend Mitchell, whom I had met when I studied shiatsu massage at the Ohashi Institute a couple of years earlier, started working on my leg. In addition to massage, Mitchell also used moxa, a Japanese heat treatment that ignites small cones made of mugwort—a root that has been used for many medical purposes, dating all the way back to the days of ancient Rome—that are placed directly on the afflicted area to alleviate congestion and pain. Finally, after only a few sessions of these combined treatments, I began to feel some semblance of relief.

But my world of pain was truly rocked the day I walked into the Long Island office of a tie-dye-worshipping ex-hippie named Harlan Sparer who practiced Directional Non-Force Technique (DNFT). Harlan looked as if he'd stepped right out of a tent at Woodstock, with his long, scraggly beard and a bunch of turquoise and copper bracelets he wore up and down his arms to ward off evil spirits. I'm not joking about the jewelry. In fact, when he saw me, Harlan told me I was emanating so much pain that he needed to put on another copper cuff for extra protection. Coming from a traditional nursing background, I thought this guy was crazy. Still, I was so desperate to find any type of relief that I was willing to give whatever he had to offer a shot.

DNFT, developed by the late Dr. Richard Van Rumpt (1904–1987), is a form of chiropractic that analyzes subluxations (misalignments of tissue both osseous and soft) through a process that involves a *challenge* and a *leg check*. The challenge, a gentle push against a body structure in a specific direction, is followed by an observation of the length of the "reactive leg" to determine whether or not that body structure is causing nerve interference. If the check is positive, then the doctor delivers a light thumb impulse in a specific direction to the body structure to adjust the misalignment. Unlike traditional chiropractic, DNFT does not involve any snap, crackle, or pop maneuvers that many people shy away from. Nor does it require an endless series of office visits. In fact, the goal of this unique practice is to resolve subluxations within as few sessions as possible.

When Harlan first performed these light touches and checks on me, I almost laughed out loud. As a traditional pre-med student, I couldn't understand how such subtle movements could possibly make any difference in my body's intense pain. But boy, was I wrong. I saw Harlan for more adjustments and, finally, started to get better. I went as a patient but was eager to learn everything I could about DNFT, while I was in chiropractic school. The results of this technique were undeniable: it really worked! And I was determined eventually to share it with as many people as I could.

Today, I am only one of about 150 DNFT practitioners in the country. Even though studies show that DNFT is the only chiropractic technique to fix low back pain in six visits, we are a tiny school of fish in a big sea of false hope and short-term fixes. Writing about the technique in the *Chiropractic Journal of Australia*, Kim B. Khauv and Christopher John concluded that as a result of it, "improvements appeared to be significant on general health functional disability, and pain intensity after an intervention of four weeks with six visits of DNFT care." Such dramatic improvements surely mean that the technique ought to be widely known and practiced.

In my practice, I see a lot of "forever." When I ask my patients how long they've had pain, swelling, digestive issues, and inflammation, "forever" is the most common answer. DNFT is the only cure I've seen to end *forever* in a very short amount of time. How does it work so well? Like every modality I use in my practice, DNFT deals with the underlying issues and not just the symptoms. Most back pain, for example, is caused by misalignments in the discs along the spine. Traditional chiropractic adjusts only the spine's vertebrae in the hopes that the discs will follow. But DNFT works directly with the discs to get to the root of the problem.

In addition, DNFT practitioners deal with each individual case as a unique situation. Traditional chiropractors have only a handful of tricks up their sleeves that they apply across the board to all of their patients. We, on the other hand, are trained like detectives to listen to our patients and search for clues to solve their particular medical mysteries. I know from my own bad experiences with doctors who aren't paying attention how important it is to listen to your patients. In some cases, I believe I actually knew better what was happening in my body. Had my doctors listened to me, I might have been spared some unnecessary treatments, including back surgery that in the end caused more damage than good.

If you picked up this book, you too most likely have been suffering from unexplained symptoms that feel as if they've lasted forever. If that's the case, then it's time to put on your own detective hat and start to ask questions. Don't just take your doctor's word for it. Your doctor may be an expert in a certain field of medicine, but only you can be an expert about your own body.

To illustrate this point, I'll give you an example of why it's important to make your own observations and assessments about your health. Last summer, a giant tree fell on the roof of my garage, requiring major reconstruction work. It was really a scary sight that I knew would cost me an arm and a leg. To figure out how much of the repairs they would cover, my insurance company plugged certain numbers into a neat little formula and came up with an amount that the company stood by as fair compensation. Unfortunately, the algorithms of their handy little system didn't take certain variables into account beyond how much waste removal would be involved, how many hours of labor they predicted the project would take, and the measurements of building materials required. Because of this, I was left paying a lot more out of pocket than my plan covered and was left with a half-repaired garage.

Similarly to the insurance company in my garage debacle, the medical world is continually shortchanging patients on the healthcare they deserve. Doctors choose to cut corners, ignore important information, and neglect to look beyond the surface of their patients' suffering. In my opinion as a healthcare professional, failing to see past numbers on perfunctory test results and images on x-rays to find answers is a form of malpractice that goes completely unchecked in this country. Doctors need to act like human beings—not test result scanners—and look their patients in the eye to find out their whole story and beyond.

Instead, what most doctors do when their patients' symptoms don't fit into neat little formulas is invent blanket diagnoses that cover a wide range of ailments. Some of the most common ones you may have heard your doctor use include the following:

- Thoracic Outlet Syndrome

    This refers to pain going down your arm or in your forearm or fingers, which may be due to a misaligned disc in your neck or T1 rib. The pain could also be caused by a number of cervical misalignments that are easily fixed, if a doctor knows what to look for. Most doctors don't and just tell patients they need to learn to live with it. In fact, the Mayo Clinic doesn't hesitate to define Thoracic Outlet Syndrome as "a group of disorders," admitting that "sometimes doctors can't determine the cause of thoracic outlet syndrome" and that "Treatment for thoracic outlet syndrome usually involves…pain relief measures." Collectively, these statements are a clear denial of their responsibility as medical professionals to get to the bottom of and actually cure the illnesses that their patients are suffering from!

- Exocrine Pancreatic Insufficiency (EPI)

    I have been treating gallbladder and pancreatic digestive issues for over twenty-five years. Now that pharmaceutical companies have created a medication for these issues (essentially the same type of natural enzymes I prescribe in my office, plus lots of preservatives and other additives), the medical world needs a whole new disease to go along with it. Note that it is not called a *disorder* or *syndrome*. It's an *insufficiency*, which just means that something is missing. According to these mainstream medical practitioners, EPI is a lack of pancreas enzyme, leading to bad digestion. But we have known that these enzymes being missing causes bad digestion for years—and the identification of this new *condition* seems to come with no attempts to identify the underlying problems that cause it!

- Patellofemoral Pain Syndrome

    This is the name given to most knee and leg pain, whether the patient is eight years old or eighty. Some doctors call it growing pains too. Others employ the terms "jumper's knee" or "runner's knee." But even for all its names, the American Academy of Family Physicians admits, "The exact cause of patellofemoral pain isn't known. However, doctors think it's related to how your leg is aligned." This syndrome is due to overuse of the knee and legs from exercise such as running or squats. All patients need to do is stretch the right amount, rest the legs, and ice and compress the area. Contrary to their doctors' opinions, they do not require medicine or surgery for this issue.

- Failed Back Surgery Syndrome (FBSS)

Back surgery for back pain could be considered malpractice, according to the *Journal of the American Medical Association* (JAMA), a highly regarded peer-reviewed medical periodical published by the American Medical Association. The reason for this is the long-term debilitating effects. For example, while back surgery can be effective at relieving sciatic leg pain, five years down the road, patients often experience terrible low back pain that significantly lowers their quality of life. The fact that there is a name given to legitimize a poor medical practice should indicate how *backward*—pun intended—the medical world can be when diagnosing patients.

You will recall reading about Failed Back Surgery Syndrome earlier in this book. As you can tell, its existence is not only evidence of the inefficacy of the widespread surgical efforts against back pain, it's also an umbrella diagnosis that physicians can hide behind when they don't know the actual cause of a patient's persistent back pain—and therefore also don't know an effective way to cure it.

As you can see from these generic names for pain, a traditional medical diagnosis is often no better than an insurance company algorithm. And the prescriptions that come with them are a cheap patch-up job to a much bigger problem.

If you know anything about me or my practice, you know that I have a low tolerance for low quality healthcare. That's why when I started to see patients who responded well enough to DNFT but were still left with inflammation, I continued to ask more questions. I realized that something else was occurring in their bodies besides misalignments. Even with nutritional supplements, these patients still experienced swelling and discomfort. Very soon, I found myself on a new search for answers.

My Quest for a Solution

Initially, I assumed that my patients' lingering inflammation was coming from their adrenal response to stress. As you may already know, stress signals the adrenal glands to produce more of the hormone cortisol, which eventually tears down connective tissue, ligaments, tendons, and discs. This inflammatory response is the reason why stressed-out people experience a lot of achiness and emotional agitation. Stress also increases the body's production of insulin, another inflammatory hormone that tells the body to store more fat. As I mentioned in the introduction, this fat-storing response is why a lot of people, by the time they reach middle age, develop love handles and muffin tops around their bellies.

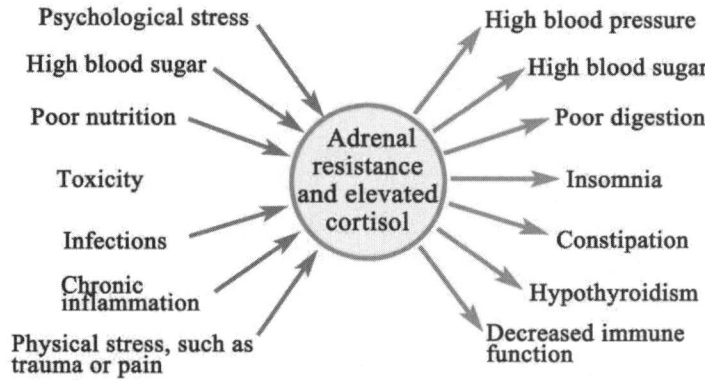

When I questioned my patients who were still experiencing inflammation after DNFT treatments, I realized that none of them had very stressful lifestyles. Nor were they necessarily retaining fat around their middles. I realized that something else besides stress hormones was keeping their lymph fluid from draining. It took an unlikely trip to far-flung South Dakota where I purchased my first lymphatic drainage machine to figure out that toxicity was the hidden cause for many people's inflammation and swelling.

The stories I heard about people's amazing recoveries from chronic pain and inflammation from sessions with this machine came at the same time I was learning about environmental toxins. From this synthesis of research and hands-on experimentation, the heart of my practice was born. In the next chapter, I will discuss in depth how our exposure to toxins is getting and keeping people in a chronic state of inflammation.

What You Need to Know

- Directional Non-Force Technique (DNFT) is the only chiropractic technique to fix low back pain in six visits.
- DNFT works directly with the spinal discs to get to the root of the problem.
- Doctors using blanket diagnoses to describe a wide range of ailments is arguably a form of malpractice that leaves patients misguided and uninformed.
- The main culprit besides stress for chronic inflammation is a high toxicity level in the body.

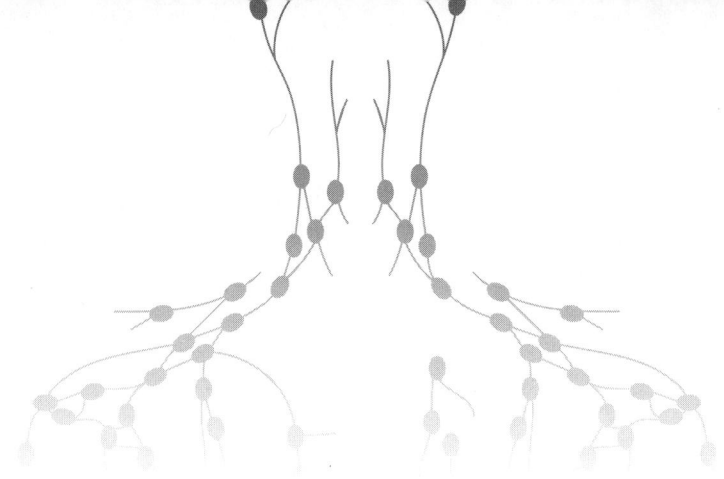

CHAPTER 5

How Toxic Are You?

Lightspring / Shutterstock

Each year, environmental pollution kills more Americans than war, violence, smoking, hunger, or natural disasters combined. From filthy air to contaminated water supplies to the toxins in our food, we are a nation in a pollution crisis. While over the past few decades, we have made progress in cleaning up the planet, the 2020 COVID-19 pandemic showed the world that we still have a long way to go. The number of seemingly-healthy Americans who died in a matter of days after infection due to lung failure and other COVID-19-related complications set off alarms that should have gone off long ago.

Still, instead of addressing toxicity levels, which is a much longer discussion, the government worked with pharmaceutical companies to control the death rates by producing multiple vaccines to mitigate the disease's symptoms. While the death rates fortunately decreased, the chemicals and pollution all around us and within our bodies remain at large.

Make no mistake: even with COVID-19 under control, we are still under siege by the toxins in our air, water, and food. More than AIDS, tuberculosis, and malaria combined, toxicity poses an invisible threat to people of all ages and walks of life. It's the leading cause of countless deadly conditions. Diphtheria, tetanus, whooping cough, diarrhea, hemolytic uremic syndrome, cholera, scarlet fever, toxic shock, and gas gangrene are all diseases related to toxins.

Not all toxins are manmade. Many are found in nature and in the foods that grow from the earth that we think are making us healthy. Foods that contain common toxins include cherry pits, apple seeds, bitter almonds, real raw cashews (before they have been steamed for resale), rhubarb leaves, raw kidney beans, and green potatoes. Here's an interesting fact about nutmeg: a little is great for taste. However, two teaspoons can make you hallucinate, or cause drowsiness, dizziness, confusion, and even seizures. These symptoms are caused by myristicin, an oil found in the spice. Another interesting tidbit is that elderberries, a great immune booster in syrup form, are full of lectin and cyanide, two substances that are poisonous when taken raw and unprocessed.

When it comes to the deadly toxicity levels we encounter on a daily basis, we are not talking about remote parts of the world, such as Sub-Saharan Africa or Asia, but our workplaces, parks, backyards, living rooms, and refrigerators. Even though the United States possesses greater pollution-monitoring technology than much of the rest of the world, we are still ignoring more than half of the 5,000 new industrial chemicals that have been developed in this country since the 1950s. In short, when it comes to the toxicity levels we are tolerating, we are not the *woke* nation that we think we are.

Lymphedema: Another Fake Diagnosis?

In my own practice, some of the highest levels of toxicity I see are found in cancer patients who come to me with lymphedema caused by post-surgery structural damage to their lymphatic systems. Sloppy mastectomy surgeons who have removed far more lymph nodes than necessary are some of the major players causing the problems. While we can blame these physicians, the real culprit of swelling and inflammation is toxicity. People have surgery every day, and many might not experience any side effects when lymph vessels and nodes are damaged or removed. As I have seen time and time again, the difference lies in the amount of toxins a patient is already harboring in their system.

A patient who comes to me with Leaky Gut Syndrome, for instance, has an immune system that has already been on overdrive due to the toxins they've ingested that are now literally leaking through their small and large intestines into the rest of their body. The antibody response to these toxins as their immune systems fight these foreign entities leaves people feeling depleted, achy, nauseous, and sick at best. The swelling and inflammation that occur beyond this point are what we know as lymphedema.

Now, what I'm about to say might sound strange, coming from a lymph specialist and the author of this book, but bear with me. Since most doctors are not addressing the toxicity that is the real, underlying cause of the disease, the term *lymphedema* is no more meaningful than the bogus diagnoses I mentioned in Chapter 2. I'm not saying the symptoms are not real. Believe me, I've seen enough grapefruit-sized swellings and witnessed enough pain in the eyes of my patients to know that their struggle is as real as the sky is blue. Yet putting a neat little label on the problem has allowed most healthcare practitioners to ignore the real cause of people's suffering: toxicity.

As I mentioned in the last chapter, a normal toxicity level is less than or equal to one. Anything above about a 1.07 indicates a growing level of toxicity that a person should start to pay careful attention to, since the numbers will only increase over time. What exactly do I mean by *pay attention*? The first step is understanding where these toxins are coming from, meaning the air we breathe, the soil we plant our fruits and vegetables in, the dairy and meat products produced in our factory farms, and the water we drink and bathe in.

The Toxins in Our Environment

Mankind has managed to drastically change the chemistry of the environment. Every year, tens of millions of pounds of chemicals are released into the underground water table beneath it. Polychlorinated biphenyls (PCBs), pesticides, fluoride, and other heavy metals are only a few examples of the smorgasbord of contaminants found in each glass of water we drink. Additionally, immeasurable amounts of chemical emissions continually are pumped into the air we breathe.

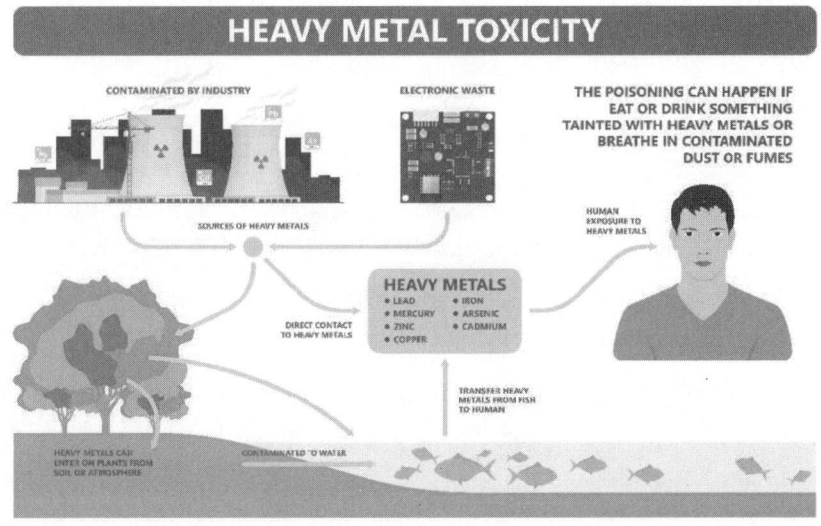

rumruay / Shutterstock

The food industry in this country is no better. Giant factories have managed to strip away most of the nutritional value of our food supply and replace it with fillers, conditioners, and artificial coloring and flavoring. Genetically modified organisms (GMOs) can be found in many food products, creating internal pollution that, when combined with predispositions, creates a scary snowball effect of disease. As is often the case with patients who come to see me with undiagnosed pain, the helpful, naturally-occurring bacteria in the intestinal tract are overtaken by this internal pollution of unnatural chemicals. As a result, free radicals, the byproduct of unhealthy bacteria, are released into circulation and negatively impact a person's overall health.

The Big Players in Toxicity

Ideally, your body's level for chemicals should be non-detectable. However, for most Americans, this is far from reality. Constant exposure to even low levels of toxins can cause dysfunction in many systems in the body. Here are just some of the big players in toxicity found in the average home in this country.

Trichloroethylene

As a rule of thumb, if you can't say it, it shouldn't go near your body, much less be inhaled or ingested. Unfortunately, this is not the case for most Americans when it comes to trichloroethylene (TCE). Before we get into what this chemical is and how it's such a big player in the world of toxicity, first, answer these questions to see why you need to care:
Do you:

1. Work close to a copy machine?
2. Drink decaffeinated coffee?
3. Use any of the following: typewriter correction fluid, rug cleaners, disinfectants, spot removers, cleaning supplies, metal degreasers?
4. Do recreational painting?

Trichloroethylene is a colorless liquid that evaporates quickly. Its two main uses are as a solvent and as a chemical used to make other chemicals. It is found especially in a common refrigerant and removers for greases, oils, fats, waxes, and tars. The textile industry uses TCE all the time to scour cotton, wool, and other fabrics. Dry cleaners use it. TCE is also a component of adhesives, lubricants, paints, varnishes, and pesticides. In a nutshell, TCE is everywhere.

Most of the TCE used in the United States is released into the atmosphere by evaporation or moves through the soil into groundwater and then up into air spaces beneath buildings where it enters the indoor air. This means that people are exposed to the chemical in a variety of ways, including TCE-contaminated air, water, food, or soil, or direct skin contact. You may also be consuming TCE-contaminated foods without even knowing it or using consumer products that contain TCE such as the ones listed above. You, your children, and your pets most likely have had direct contact with soil that contains TCE as well.

Most of the trichloroethylene that you breathe goes straight into your bloodstream and organs. A small amount can also enter your bloodstream through your skin. The TCE found in foods ranges in concentration levels between 2 and 100 ppb, and ends up—you guessed it—in your bloodstream. Once the TCE is in your blood, your liver changes it to other chemicals that end up getting stored in your fat and eventually come out in your breath and urine.

To give you an idea of how concentrated amounts of it can affect you, TCE was once used as an anesthetic for surgery. Depending on the amount and length of exposure, people exposed to it experience symptoms ranging from sleepiness, dizziness, and headaches to nerve damage, irregular heartbeats, liver and kidney damage, coma, and death. People exposed to TCE in the workplace may develop autoimmune diseases, a decrease in sex drive, sperm quality, and reproductive hormone levels. There is strong evidence that TSE causes kidney cancer, liver cancer, and malignant lymphoma. It has also caused leukemia, testicular cancer, and lung tumors in rats and mice.

Unfortunately, because of its pervasiveness, there is very little you can do personally to lower your exposure to TCE. It's up to the government to enforce stricter regulations to decrease manufacturers' use of TCE and other toxic chemicals. Regulations (enforceable by laws) and recommendations (not enforceable by laws) are expressed as "not-to-exceed" levels. How does the government decide what levels of toxins are okay to release? The regulations and recommendations are based on levels that affect animals in laboratory tests; levels are then adjusted to help protect humans. These not-to-exceed levels can be arbitrary and differ greatly among federal organizations. According to the Agency for Toxic Substances and Disease Registry, when testing toxic chemicals, "Different organizations use different exposure times (e.g., an eight-hour workday or a twenty-four-hour day), different animal studies, or emphasize some factors over others, depending on their mission."

Formaldehyde

If you thought this stuff was only used to preserve dead things, think again. Formaldehyde is one of the most common toxic chemicals around. If you answer yes to any of these questions, then you risk exposure to it.

Do you:

1. Wear dry-cleaned clothing?
2. Have foam wall insulation, particleboard, chipboard, or interior plywood in your home?
3. Have foam cushions or mattresses?
4. Work in a laboratory?
5. Have recently-purchased carpets?
6. Use wax or polish on your floor?
7. Have a home with plaster, stucco, or concrete?
8. Have draperies?
9. Have a wood-burning stove?
10. Smoke in your home?
11. Use nail polish remover?
12. Use fingernail hardeners?
13. Have fireproof material in your home?

The list goes on. The reality is, there are small amounts of formaldehyde in nearly every home in this country. Since tobacco smoke contains formaldehyde, if someone in your home smokes, their cigarette, cigar, or pipe smoke may be the greatest source of formaldehyde there. Homes with new products or construction also contain higher amounts of formaldehyde. In particular, you'll find large amounts in plywood, particleboard, and laminate flooring, as well as permanent-press fabrics such as those used for curtains or upholstery. Household products such as glues, paints, pesticides, cosmetics, and detergents also contain formaldehyde in varying amounts. In addition, homes built after 1990 tend to have better insulation, and, therefore, can keep formaldehyde lingering inside longer.

Formaldehyde, like most airborne toxins, can cause breathing problems or irritation to the eyes, nose, throat, or skin. Children, older adults, and people with asthma and other breathing problems are affected the most by these symptoms. Breathing in very high levels of formaldehyde over many years has been linked to rare nose and throat cancers.

So what can you do to limit your exposure to formaldehyde? Encourage ventilation by opening your windows for a few minutes every day and installing exhaust fans. Make your home smoke free and keep the indoor temperature and humidity at the lowest comfortable setting. For future purchases, you should choose home products with low or no formaldehyde. Look for furniture and flooring without urea-formaldehyde (UF) glues, pressed-wood products labeled ULEF for ultra-low emitting formaldehyde or NAF for no added formaldehyde requirements, and insulation that does not contain UF foam. You should also wash permanent-press clothing before wearing it and let new products release formaldehyde outside of your living space before using them inside.

Pesticides and Herbicides

Everyone loves a pretty, green lawn free from weeds and bugs. But that pretty picture can come at an ugly cost when it comes to toxicity. Pesticides and herbicides make their way into our everyday lives and accumulate in our bodies, causing a huge amount of health problems. How prevalent are they? Answer these questions, and you'll get a good idea.

Do you:

1. Use weed killer on your lawn?
2. Drink tap water?
3. Have mothballs in your closet?
4. Get bothered by gasoline fumes?
5. Eat store-bought meat?

In the last half of the last century, worldwide pesticide and herbicide production increased at a rate of about 11 percent per year, from 0.2 million tons in the 1950s to more than 5 million tons in 2000. Now, here's the kicker: only 0.1 percent of applied pesticides reach the target pests, leaving 99.9 percent of the chemicals to linger in the environment. Keep reading to decide for yourself whether the benefits of such high levels of pesticide usage outweigh the risks.

Pesticides and herbicides have long been associated with short- and long-term effects on human health, such as elevated cancer risks, including melanoma, and disruptions of the body's reproductive, immune, endocrine, and nervous systems. Most of our exposure to these chemicals comes from absorption through the skin or by ingestion. However, they are also in the water we drink and the air we breathe. Many respiratory pathologies are also related to occupational exposure to pesticides, including asthma, chronic obstructive pulmonary disease (COPD), and lung cancer.

Despite these health risks, very few people are aware of their exposure to dangerous pesticides and herbicides before it's too late, and they are seriously ill. What's more, the big companies that produce them have very little motivation to stop until they are fiscally at risk. Take the glyphosate-based weed killer, a widely-used herbicide, that has been found to stimulate non-Hodgkin's lymphoma and breast cancer growth via estrogen receptors. The company that owns this weed killer—and yes, the same company that makes those bottles of pills in your medicine cabinet—has recently settled most of the lawsuits brought against it by plaintiffs alleging that the glyphosate used in the weed killer causes cancer. The original company that developed glyphosate has long maintained that the weed killer does not cause cancer. Still, in July 2021, the new company said it would remove all glyphosate-based herbicides from the US consumer market by 2023.

In addition to glyphosate, malathion, parathion, and dimethoate, all found in pesticides, are known to disrupt endocrine function and are associated with effects on cholinesterase enzymes, a decrease in insulin secretion, disruption of normal cellular metabolism of proteins, carbohydrates, and fats, genotoxic effects, and effects on mitochondrial function, which can lead to problems with the nervous and endocrine systems. Besides the health risks they pose for people, these herbicides and pesticides are endangering the entire food chain, including birds, mammals, fish, and honeybees.

What can you do about your exposure to these chemicals? More than you think. Here are just a few practices you can implement today to limit your intake:

- Eat a *variety* of fruits and vegetables to minimize the potential of increased exposure to a single pesticide.
- Thoroughly wash all produce under running water, rather than soaking or dunking it, even if it's labeled organic or has a peel.
- Dry produce with a clean cloth towel or paper towel.
- Scrub firm fruits and vegetables, such as melons and root vegetables.
- Discard the outer layer of leafy vegetables, such as lettuce or cabbage.
- Peel fruits and vegetables when possible.
- Trim fat and skin from meat, poultry, and fish to minimize accumulated pesticide residue.

In addition to getting rid of the pesticides, you can also consider growing your own pesticide-free garden or participating in a community garden. Another option is shopping at your local farmers market, where you can speak directly to the farmers about their pesticide use practices before buying their food.

Other Volatile Organic Compounds

Volatile organic compounds (VOCs) are chemicals that have a high vapor pressure and low water solubility, properties that help them stick around in our air and water supplies for a long time. Many VOCs, such as formaldehyde, which I already talked about, are used in the manufacture of paints, pharmaceuticals, and refrigerants. VOCs typically are industrial solvents, such as trichloroethylene, fuel oxygenates, or by-products produced by chlorination in water treatment, such as chloroform. VOCs are also components of petroleum fuels, hydraulic fluids, paint thinners, and dry-cleaning agents. Let's see how many of them you might be exposed to each day.

Do you:

1. Use cleaning solvents?
2. Have soft vinyl floors?
3. Handle propane or butane?
4. Get your clothes dry-cleaned?
5. Work close to a laser printer?
6. Use moth balls?
7. Have nylon carpet?
8. Use air fresheners?

VOCs are found up to ten times more frequently indoors than outdoors. They are found in thousands of everyday products, including paints and lacquers, paint strippers, cleaning supplies, pesticides, building materials and furnishings, office equipment such as copiers and printers, correction fluids and carbonless copy paper, glues, permanent markers, and photographic solutions. VOCs have a variety of short- and long-term health effects, including the following: eye, nose, and throat irritation, headaches, nausea, loss of coordination, and damage to the liver, kidneys, and central nervous system. They can also cause cancer.

Since no federally-enforceable standards have been set for VOCs in non-industrial settings, it's up to you and your family to limit your exposure to them. Here are some things you can do to protect yourself from these harmful chemicals:

- Increase ventilation when using products that emit VOCs.
- Do not store opened containers of unused paints and similar materials in living spaces.
- Use integrated pest management techniques to reduce the need for pesticides.
- Use household products according to the manufacturer's directions.
- If dry-cleaned goods have a strong chemical odor when you pick them up, do not accept them until they have been properly dried.
- Make sure you provide plenty of fresh air when using these products.
- Throw away unused or little-used containers safely in designated dumps.
- Buy products containing VOCs only in quantities that you will use soon.

Phenols

Phenols are everywhere, from household cleaners and mouthwash to Styrofoam cups and automotive products. But did you know that these toxic chemicals are also found in throat lozenges and skin ointments we give our kids? Here are just some of the items containing phenols that are in your home right now: household cleaners, nasal spray, cough syrup, adhesive tape, scented deodorant, newsprint, epoxy, some mouthwash, mildew cleaners, air fresheners, perfumes, hard saucepan handles, hair spray, and air sanitizers.

Phenols are highly irritating to the skin, eyes, and mucous membranes after inhalation or dermal exposures, and are considered to be very toxic to humans via oral exposure. Symptoms from long-term exposure include progressive weight loss, diarrhea, vertigo, salivation, a dark coloration of the urine, blood and liver effects, anorexia, irregular breathing, muscle weakness and tremors, loss of coordination, convulsions, coma, and respiratory arrest at lethal doses. Phenols are mainly incorporated in the production of phenolic resins, which are used in the plywood, construction, automotive, and appliance industries. Phenol is also used as a slimicide, a disinfectant, and in medicinal products such as ear and nose drops, throat lozenges, and mouthwashes.

The best thing you can do to limit your exposure to phenols is not to buy products containing them. Read labels! And if you have to purchase products containing phenol, limit the quantity and use of them. Store them either in a basement or garage where you spend little time. Research alternative, non-toxic products and invest in those instead.

BPA

Most people know that plastic water bottles are terrible for the environment. But did you know that they also seep out lots of chemicals into the water they contain? Research suggests that all plastics may seep chemicals if they're scratched or heated. Research also strongly suggests that at certain exposure levels, some of these chemicals, such as bisphenol A (BPA), may cause cancer. BPA is a weak synthetic estrogen found in many rigid plastic products, including most water bottles and food packaging.

BPA's estrogen-like activity makes it a hormone disruptor, like many other chemicals in plastics. Hormone disruptors can affect how estrogen and other hormones act in the body by blocking them or mimicking them, thus throwing off the body's hormonal balance and causing hormone-receptor-positive breast cancer. BPA also affects brain development in the womb, causing female babies to show signs of hyperactivity, anxiety, and depression.

Here are some steps you can take to reduce your exposure to BPA and other hormone disruptors:

- Carry your own glass, steel, or ceramic water bottle filled with filtered tap water.
- Use baby bottles with labels that say "BPA free."
- Try to use as little plastic as possible, especially if you're pregnant.
- To reduce your exposure to other chemicals in plastics, do the following:
- Don't cook food in plastic containers or use roasting/steaming bags, which contain plastic residues that may seep into food when heated.
- Use glass, porcelain, enamel-covered metal, or stainless-steel pots, pans, and containers for food and beverages whenever possible.
- Plastics with recycling symbol 2, 4, and 5 are generally considered safe to use. Plastics with recycling symbol 7 are safe to use as long as they also say "PLA" or have a leaf symbol on them. The recycling symbol number is the code that shows what type of plastic was used to make the product.
- Recycling symbol 1 is also safe to use but shouldn't be used more than once (i.e., don't refill those store-bought water bottles). Keep all plastic containers out of the heat and sun.

Now that you are more aware of the world of toxicity we live in, in the next chapter, I will show you some simple ways you can protect yourself and your family from too much exposure to toxins.

What You Need to Know

- The number of deaths caused by the 2020 COVID-19 pandemic revealed the high toxicity levels of many Americans.
- The term *lymphedema* allows doctors to ignore patients' underlying toxicity that causes their swelling and inflammation.
- Our water supply contains PCBs, pesticides, fluoride, and other heavy metals that our bodies are absorbing each day.

- Trichloroethylene (TCE), formaldehyde, phenols, and other VOCs are all common toxins found in our soil, groundwater, and indoor air spaces.
- These chemicals can cause many harmful side effects such as nerve damage, irregular heartbeats, liver and kidney damage, coma, possible kidney cancer, liver cancer, malignant lymphoma, asthma, rare nose and throat cancers, irregular breathing, muscle weakness and tremors, loss of coordination, convulsions, and respiratory arrest.
- Pesticide and herbicide production has increased at a rate of about 11 percent per year, from 0.2 million tons in the 1950s to more than 5 million tons in 2000.
- Exposure to pesticides and herbicides can result in elevated cancer risks, a disruption of the body's reproductive, immune, endocrine, and nervous systems, and respiratory pathologies such as asthma, chronic obstructive pulmonary disease (COPD), and lung cancer.
- BPA is a synthetic estrogen found in many rigid plastic products, including most water bottles and food packaging.
- BPA throws off the body's hormonal balance and causes hormone-receptor-positive breast cancer.
- You can limit your exposure to BPA by carrying your own glass, steel, or ceramic water bottle, not cooking food in plastic containers, paying attention to the recycling numbers on plastics, and not reusing plastics that contain the recycling number 1.

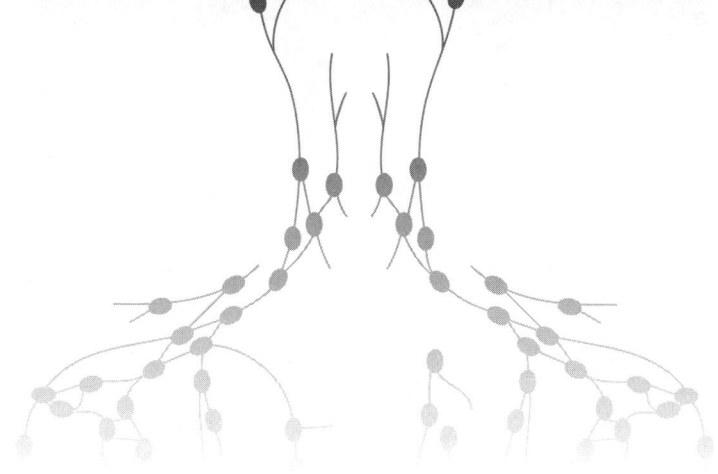

CHAPTER 6

Simple Steps to Detox

As you've already seen, toxins are all around us. The winning combination of lymph drainage and metabolic detoxification, which I will discuss in Chapter 7, is the best way to change the amount of toxins lymphedema patients absorb and hold on to. Much of what we can do to detoxify has to do with lifestyle and diet. Here are just a few steps to take that can make all the difference in a person's toxicity levels.

Exercise

A good workout sounds as if it would help get rid of toxins through the sweat that comes out of your pores. However, while exercise does play a huge role in the detoxification process, the benefits don't have much to do with sweat. In fact, toxins released through perspiration (if any) are likely an overall drop in the bucket compared to what's eliminated via urine and stool. You see, sweat's main function is thermoregulation, not detoxing. As I've already discussed, the main way your body removes toxins is through the liver and kidneys, which process them and expel them through urine and stool.

Still, exercise does play an important role in the detoxification process. The good news is you don't have to run a marathon or do a million burpees or squats to make exercise count for you. By simply moving your body for extended amounts of times through leisure activities such as walking, biking, gardening, doing housework, and kayaking, you'll be doing your body good. Our bodies were designed to move, not sit in front of a computer screen all day. Unfortunately, the way the modern world is set up, people are moving a lot less than they did a hundred years ago. Not only this, but we are also exposed to many more toxins, as I've just discussed.

So how does exercise help? First, it increases circulation. All cells are water-based and bathed in extracellular fluid that carries toxins away to the liver and kidneys for disposal. It makes sense that the more you move, the less that fluid will stagnate and allow toxins to build up in your body.

The second way exercise can help with detoxification is by lowering inflammation. Besides the many toxins we take in from the outside world, chronic inflammation is itself toxic to our cells and produces an overabundance of natural "toxins." Normally, free radicals, which are the byproducts of cells as they make energy, are used to break down cellular waste that is then carried away through the lymphatic system and filtered through the lymph nodes. However, when cells are stressed by environmental toxins, they produce more waste, leading to a flood of excess free radicals. These free radicals begin to then break down healthy tissue at increasing rates, leading to more waste. It's a vicious cycle that contributes to chronic illness and accelerates the general aging process.

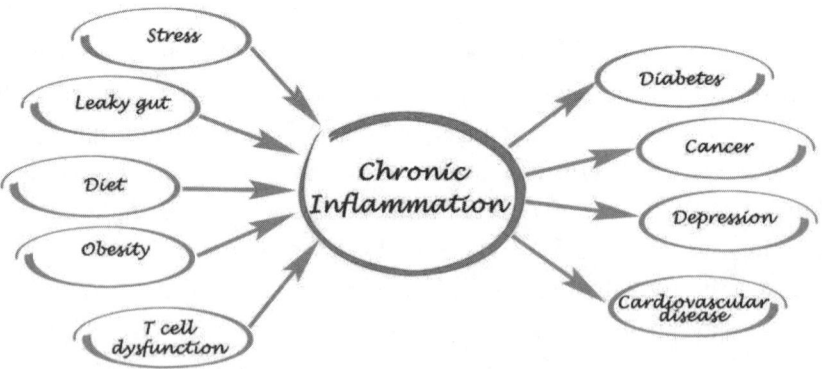

In short, chronic inflammation is a reaction to the whole plumbing system of the body becoming overwhelmed and getting backed up. Your cells are so polluted that your system is collecting waste at a faster pace than it can get rid of it. Physical activity plays an important two-part role in detoxification because it helps both reduce inflammation *and* clear lymphatic congestion.

Lastly, exercising is an excellent way to get your bowels moving to excrete toxins. It decreases the time it takes for food to move through the large intestine, helping you to avoid hard stools that are more difficult to pass. Aerobic exercise accelerates your breathing and heart rate to help stimulate the natural contraction of intestinal muscles, an activity that will also help move stools out quickly.

Weight

Fat tissue is the perfect place for toxins to hide. Overweight people tend to have higher body burdens of common environmental toxins. They experience more fatigue than other people due to the extra weight they carry, as well as a slowed metabolism. Their bodies might appear robust, but their cell membranes often are shrinking, as the extra fat cells block cell receptors, making it harder to get rid of toxins. Exercise, along with a healthy diet, helps reduce fat tissue and keep your weight and, potentially, the amount of toxins in your system in check.

Diet

Eating "clean" means cooking with whole foods and staying away from any food that is frozen or processed. Purchasing organic products whenever possible will also eliminate a large number of chemicals that enter the body. Also, eating colorful fruits and vegetables is important. Avoiding inflammatory foods is also a must. In Chapter 10, I will cover all of the dietary guidelines that will help your body's natural detoxification process.

Water

We know that water flushes toxins and waste from the body and transports nutrients to where they are needed. Without water, the contents of your colon can dry out and get stuck, eventually causing constipation. Water is a natural lubricant that softens stool and promotes the evacuation of the bowels.

Water is also essential to the kidneys' ability to filter and remove waste products from the blood, eliminate toxic substances in the urine, and receive water-soluble toxins from the liver for processing. The kidneys filter voluminous amounts of blood each day and so maintain the body's water balance and excretion of toxins and excess fluid through the bladder. Daily fluid intake is essential for our bodies to function efficiently.

How much water should you drink? Eight to ten glasses a day. Most people only get about half of this and don't realize that they are water-deprived most of the day. Here's a clue to know if you aren't getting enough water: if you are thirsty, then you are already dehydrated. Don't wait to want water. Make it a habit before thirst even sets in. And make sure the water is as pure as it can be, according to my recommendations earlier in the chapter.

Besides drinking, there are many other ways water can help you remove toxins. Water is also great at detoxing in baths and saunas. Hot water increases blood flow and capillary action near the surface of the skin, causing faster release of toxins. Add a little bentonite clay, a powdery substance you can buy at the health food store, to your bath to aid in detoxification.

Medication

Biotoxicity from too much medication in a person's system can cause an overuse of cytochrome P450 (CYP450) enzymes that are normally used to fight free radicals, which contribute to inflammation and disease. CYP450 are also essential for the production of cholesterol, steroids, prostacyclins, and thromboxane A2. They are necessary for the detoxification of foreign chemicals and the metabolism of drugs. Drugs that interfere with CYP450 function are referred to as CYP450 inhibitors. If CYP enzymes are not active enough, medications can stay for longer time in our body, leading to toxicity. Other factors that can influence CYP450 enzyme activity include drinking grapefruit juice, eating charcoal-grilled foods, and smoking.

I try to reduce patients' load of outside medications previously prescribed by doctors, which contain many unnecessary ingredients that also pollute their systems. These include heavy insulin for diabetes and statins prescribed for high cholesterol. Many physicians are quick to prescribe a medication for every symptom, but not every symptom requires a medication. In addition, patients often see multiple doctors who do not communicate with one another and so end up prescribing similar drugs, which, when combined, can reach toxic levels.

Make sure you communicate with all of your physicians about the different medications you are taking and their dosages. Electronic medical records will also help close the communications gap among doctors. To avoid drug toxicity, patients should be proactive by keeping a careful record of which drugs they're taking—including over-the-counter medications—and bringing that list to every doctor visit.

Patients should also read the safety inserts that come with their medication—*before* taking it.

What Victoria says...

Dr. Loretta is like a family secret—but one that we want to share with the world! She gave my mom back her life when doctors told her that the only way she'd be able to have a semi-normal existence was if she had back surgery. At that point, my mother was in chronic pain, wasn't sleeping, and was on constant pain medication that was creating toxicity throughout her body. Today, she is a pain-free, fit woman, full of life and happy to be a grandmother!

I myself started seeing Dr. Loretta when I was pregnant and had terrible morning sickness. The pregnancy was compressing my organs and creating a lot of pain, which her adjustments helped alleviate tremendously.

Years later, my autoimmune issue flared up, starting as a sinus infection and escalating so that my entire left side from head to toe was out of balance and not draining properly. Additionally, I couldn't hold food down and had debilitating migraines. Dr. Loretta did Lymph-Biologics™ on my neck and spine. After one session, for the first time in months, I could move my head without excruciating pain. And I was able to get rid of my sinus infection without using antibiotics.

My children, one a gymnast, the other a professional dancer, have seen her for different aches and pains as well. What a priceless gift to our family!

What You Need to Know

- Regular exercise helps circulate the extracellular fluid that carries toxins away to the liver and kidneys for disposal.
- Exercise also reduces inflammation and clears lymphatic congestion.
- Extra fat cells block cell receptors, making it harder to get rid of toxins.
- Eating whole, unprocessed, organic foods greatly reduces the number of toxins we consume each day.
- Water is also essential to the kidneys' ability to filter and remove waste products from the blood, eliminate toxic substances in the urine, and receive water-soluble toxins from the liver for processing.
- A good amount of water to drink every day is eight to ten glasses.
- Biotoxicity caused by medications can create a vicious cycle by increasing the presence of free radicals and creating hormonal imbalances that might require even more medications.
- Make sure you keep a careful record of your medications and dosages and share it with all of your physicians.

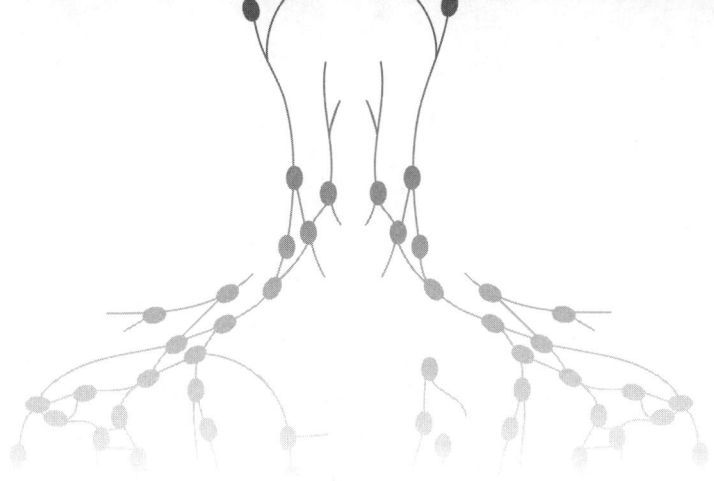

CHAPTER 7

Lymph-Biologics™ and Metabolic Detox: The Winning Combo that Works

Back when I was in nursing school, health practitioners weren't talking about using tests such as urine or saliva analyses to fine-tune diagnoses. No one mentioned the possibilities of hidden parasites or worms that might be causing chronic discomfort. And no one considered modalities such as muscle or homeopathic testing as viable options for diagnosis. Only when I embarked on my own healing journey for chronic lower back pain and TMJ did I begin to step off the narrow path of textbook diagnoses and overly-prescribed medications and start to discover alternative methods of testing and treatment that actually worked.

Out of desperation to find relief, I visited Tim Guilford, an alternative medicine doctor, and Doug Mosher, a DNFT chiropractor, who both shared an office in San Jose, California. At the time, Tim told me about an infant he'd been treating for multiple seizures that traditional doctors could not find a solution for. This baby had up to 180 seizures a day. After a thorough workup that included a saliva and urine analysis and muscle testing, Tim had found parasites in the young patient's cerebral spinal fluid that traditional testing by the infant's pediatrician had been unable to discover. Because of his great detective work, Tim was able to treat the baby and cure her of her life-threatening seizures.

A few years later, when I started treating my own patients for inexplicable ailments, I already knew from watching Tim and Doug that most musculoskeletal problems originally arose from issues far less visible to the eye than simple misalignments. My hypothesis was proven correct again and again whenever patients came to me with inflamed discs and a lot of pain, and Lymph-Biologics™, my trademarked Lymph-Drainage technique, provided them with instant, lasting relief.

How and Why Lymph-Biologics™ Works

If you are reading this book, chances are you are looking for a new, better approach to your chronic health issues, which may include ongoing, perhaps inexplicable, pain. Until now, you might not have known where to turn. Maybe you're confused by generic diagnoses or conflicting prescriptions. Or you're intimidated by misinformation that pegs alternative pain treatments as unproven or harmful. You may lack the financial resources to pursue treatments that are not covered by your insurance. The bottom line is you deserve to know what is happening to your body. You also deserve viable solutions that will provide permanent relief.

Every day, I talk to people all over the world like you who are suffering from lymphedema. Some have been suffering for only a few weeks; others for several years. Whether the swelling is caused by mechanical damage, such as an injury or surgery, or by something more insidious, all of these people have one thing in common: toxicity. In addition to lymphedema, many of my patients are suffering from all sorts of diseases such as cancers, diabetes, and auto-immune disorders, yet none of their doctors has spoken to them about their toxicity levels or about changing their diets and lifestyles. Those patients who are somewhat aware of their toxicity levels are often shot down by their doctors when they suggest toxins as a possible cause for their issues. The reason for this is unclear to me, but, knowing what I know about western medical practices, I am quite certain it has something to do with cost effectiveness and the amount of effort a holistic approach to patients' health would entail.

Unfortunately, since toxicity is the cause of long-term lymphedema, it is also the reason the disease lingers. Acknowledging this simple fact is the first step toward healing. Yet, instead of treating the underlying cause of toxicity, most doctors employ modalities that provide short-term relief, if any.

I mentioned lymphatic massage earlier as an important tool against lymph pain that is often overlooked. But the practice itself is not a true cure. It only pushes the fluid from point A to point B. It may be able to reduce the size of the swelling as well as some of the pain that comes from the heaviness and fullness of the involved area, but that is where its effectiveness ends. Lymphatic massage alone cannot remove the toxins in your body, nor can it open the blocked or stagnant ducts that handle the flow of the lymphatic fluid, so it can leave the body.

Lymph-Biologics™ is the only proven technique that actually resolves inflammation and swelling of the lymphatic system. It utilizes a three-part light wave that penetrates deep into the lymphatic system and causes a sympathetic response, which is the body's normal, healthy reaction to danger. If you picture the flight-or-flight response of an animal facing a predator, you'll get a better idea of some of the involuntary ways our bodies' natural intelligence kicks in. Some examples of sympathetic responses our bodies experience on a regular basis include an accelerated heart rate, the widening of bronchial passages, a constriction of blood vessels, pupillary dilation, piloerection (goosebumps), raised blood pressure, and perspiration.

The sympathetic response stimulated by Lymph-Biologics™ is caused by peristalsis, a series of wave-like muscle contractions that move the lymph fluid through the body's tissue. This cathartic movement is performed simultaneously with the employment of an electrostatic field that removes the toxins that are blocking the lymph fluid from draining properly.

Before I go into more detail about the mechanics of Lymph-Biologics™, let's first review the functions of the lymphatic system and why lymph fluid is vital to your body's health. When you sprain an ankle, and that part of your body swells like an orange or a grapefruit, that's your body's normal Immune System Response to an injury. The lymph fluid is actually protecting the area from any further injury. This lymph response is the innate intelligence of your body that acts without your having to even think about what to do. If that swelling subsides within four to six weeks, you know that your body is operating at normal capacity. If, however, eight weeks later, the swelling is still there, the medical world recognizes this as an abnormal response that is now called lymphedema.

Lymphedema occurs when toxins do not allow the lymph fluid to drain from internal tissue. By using Lymph-Biologics™ on my patients, I can remove tightness, heavy sensation, fullness, swelling, inflammation, and pain in any area of the body I work on. Whether it's a foot, the neck area, the breasts, or an arm, Lymph-Biologics™ works because it doesn't just move the lymph fluid around the way normal lymphatic massage does. It actually eliminates it from the tissues and deals with the root cause of blockage.

Just how does it do this? The machine's glass tubes contain the noble gases argon, xenon, and krypton that are ionized (charged) by a high voltage field set off at a low current (amperage). The therapy heads emit various fields of energy when touched to the skin similar to a static spark people pick up from walking across carpeting in dry climatic conditions that comes out as a mini *zap* when touching a ground surface. (Of course, my clients don't ever feel this spark because they are already electrically grounded.) Next, the electrostatic field, called the Dynamic Pulsed Ionic Field, is driven by a circuit that pulses it in a moving cycle, within a specific range of frequencies under 1000 Hz. This process keeps a patient's body from adapting to the energy over time and also provides richer fields of frequencies that the body can use.

Unlike other forms of electrotherapy such as ultrasound or diathermy that can heat the body's tissues and cause radiofrequency (RF) burns, the energies of Lymph-Biologics™ are non-thermal and safe. Vibrational lymphatic therapy in general improves the entire circulation system by allowing the body to release toxins and accumulated fluid and proteins between the cells. A series of treatments over time improves a patient's lymphatic flow and accelerates detoxification of their tissues. This leads to more vitality, relief of aches and pains, faster recovery from surgeries, and healing from tender cystic breasts and chronic diseases such as cancer and asthma.

Most importantly, Lymph-Biologics™ helps rid the body of all kinds of environmental toxins by increasing fluid volume and exchange at the site of the lymphatic vessels as well as interstitial spaces between cells. It also facilitates the transportation of immune functions into the cells, organs, and white blood cells. But what happens when too many toxins are attaching to the cells to allow for the flow of fluids that are so essential to good health? Metabolic detoxification in conjunction with Lymph-Biologics™ is the only way to ensure that the pathways of the lymphatic system will be free from blockage.

Why Detoxification is Essential for Good Health

As I say to all my new patients, the question isn't *Are you toxic?* It's *How toxic are you?* The main purpose of a cell test—which I will go into greater detail about in the next chapter—is to discover the answer to this question. Like a snapshot of a patient's tissue and fluid compartments, it allows me to see the toxicity levels that increase nutritional deficiencies and cause premature aging.

A lack of proper nutrients in a patient's diet is only one factor contributing to the accelerated aging of cells. Cell receptors can get blocked by toxins and lose their ability to usher in important nutrients and hormones that the cells require for energy. Excess body fat can block these receptors as well, resulting in what's called insulin resistance, which eventually leads to type 2 diabetes. Besides obesity, high levels of stress can also be the main cause of insulin resistance. As you will learn more about later in the book, stress causes an entire chemical cascade to occur that compromises your cellular profile. The resulting effects put a person at risk of not only type 2 diabetes, but also coronary artery disease, high cholesterol, metabolic syndrome, stroke, and cancer.

In short, shrinking cell membranes and hindered cell receptors are byproducts of high toxicity levels. To determine these levels, I use an algorithm based on the mass of lymph, organs, and cartilage versus how much fluid exists outside the cell. In a healthy body with normal levels of toxicity, those two numbers create a percentage that should be less than or equal to 1. Any amount over this threshold reveals how high a patient's toxicity is. How high is high? Anywhere between 1.07—which I consider borderline—to 1.20—which I consider pretty up there—is the range of toxicity for most people who walk into my office. However, every once and a while, I get a cancer patient whose number reads off the charts.

Gloria, a 36-year-old woman with stage-four colon cancer, who had received heavy-duty chemotherapy and radiation treatments that scored her a whopping 1.79, was one such patient. Once I assign a number to a patient's toxicity level, the next step is to investigate what specific toxins are contaminating their system. Figuring out what type of toxicity is in the body can be challenging, considering the myriad of harmful substances we are exposed to on a daily basis. Additional testing of urine can determine liver function and the presence of heavy metals. A stool test can establish whether a patient is suffering from Leaky Gut Syndrome, a condition that certainly contributes to toxicity.

Don't get me wrong: metabolic detoxification programs can work exceptionally well on their own. Detoxification methods of healing have been used for thousands of years by various cultures throughout the world. Fasting, one of the oldest therapeutic practices in medicine, was recommended by Hippocrates, the ancient Greek known as the "Father of Western Medicine." Ayurvedic medicine, a nutritional healing system originating in India, utilizes detoxification to prevent and treat a great number of illnesses.

Fake Versus Real Detox

Today, detoxification has become one of the great cornerstones of alternative medicine. Even conventional medicine cannot deny that toxic environmental factors play a significant role in poor health. We know that after the terrorist attacks of 9/11, cases of sixty-eight distinct cancers showed up in people who were exposed to the toxins released from the fallen buildings. Now, 75 percent of the people who worked on the debris piles for months after 9/11 happened are sick with some form of cancer. This type of exposure to toxicity, however, is not limited to tragedies such as what happened on 9/11. Every day, similar toxins are in the air we breathe, the water we drink, and the food we eat. We are all vulnerable to them, yet not all of us can rid our bodies of them. Instead, toxins accumulate inside our lymphatic tissue and cause all kinds of autoimmune diseases and cancers.

If only a simple cleanse protocol could get rid of toxicity in the body—but it can't. Cleanses are a multi-billion-dollar industry that do very little for the body since they bypass the liver, kidneys, and lymphatic system. For the same reason, even though some cleanses may do something for the digestive tract, they cannot be considered metabolic detoxification.

The Cleveland Clinic cited Dr. Kate Patton as saying that, while there is no conclusive evidence that the average person would derive any benefit from a cleanse that involved skipping certain solid foods (or all of them), the exceptions are those people who have specific digestive disorders, including Crohn's disease and gastroparesis. In fact, most cleanses, such as a water fast, a juice fast, or what's called a *master cleanse*, a ten-to-forty-day juice cleanse, are not much more than a temporary caloric deficit that temporarily clears out the small and large intestine.

These cleanses throw the body into a ketonic state so that it burns more fat as energy. In this state, people develop very unpleasant side effects such as lightheadedness and giddiness, head- and body aches, and breath that smells and tastes as bad as horse manure for several days. Although the numbers on the scale will decrease, they are only doing so from the body's loss of water weight, not fat. As I've said, a cleanse does nothing for your liver, kidneys, and lymphatic system to remove actual toxins. People spend a fortune on cleanses, but they simply do not work. In fact, according to the U.S. Department of Health and Human Services, "A 2015 review concluded that there was no compelling research to support the use of 'detox' diets for weight management or eliminating toxins from the body." The fact is that the places where toxins do the most harm to our body are not anywhere in our digestive system—which means we need to use other tools to fight these toxic substances and the harm they inflict.

Some people get very sick when they try these trendy detox regimens. Among common symptoms are severe headaches, vomiting, and feeling just awful in general. This reaction can happen even to the average healthy person who decides to do a detox or cleanse diet. According to information published by Harvard Medical School, the diets used on cleanses don't include sufficient amounts of proteins, fatty acids, and other essential nutrients. People often get as few as 600 calories on these diets—and all from carbohydrates. But that's not what all. The laxative effect of the diet can make you dehydrated—which can lead immediately to headaches and then to a whole host of medical ill effects. Low calorie content and dehydration alone will leave people feeling weak and displaying signs of low energy, low blood sugar, muscle aches, fatigue, dizziness, and nausea. It can make you have too few electrolytes, which are essential to basic body functions (and they're what we are trying so hard to replenish when we chug those big bottles of Gatorade!). The after-effects of the diet can stop your bowels from functioning normally. Even worse, the cleanse can actually "cleanse" out useful intestinal flora, which perform good functions in making sure our digestive system functions.

Some people take the harmful effects of cleanses to the next level by making a cleanse a monthly ritual. These people may develop a condition that can be as scary as its name: metabolic acidosis. That means that the very acid-base balance of your body has become disrupted, and your blood is now too acidic for your body. A severe case can mean coma and even death. Not something to play around with!

But often the worst effects of these cleanses are reserved for people who are toxic already. Even starting to burn fat and releasing all that toxicity is dangerous. Of course, as you might have guessed, a 600-calorie diet may lead to weight loss. But weight loss is not always a good thing. Why? The body uses our fat cells for storing toxins. When we lose weight, that fat is eliminated and the toxin stores in it are released into the body. If you lose weight rapidly, that's a lot of toxins for the liver to handle, and a huge shock to the system. In one study, Korean researchers from Kyungpook National University, Daegu, published in the *International Journal of Obesity* findings that showed that losing weight rapidly released POPs into the blood. What are POPs? (Hint: I'm not talking about Coke and Pepsi.) POPs is short for "persistent organic pollutants"—another kind of toxin that causes organ damage and has been linked to disruption of the endocrine, reproductive, and immune systems; dementia, and cancer.

Another study, which researchers from the Johns Hopkins Bloomberg School of Public Health published in *Obesity*, found that after rapid weight loss, people experienced "rises in the bloodstream levels of environmental toxicants that are known to be stored long term in fat, including polychlorinated biphenyls, organochlorine pesticides, and polybrominated diphenyl ethers." The last on the list is better known as "PCBs," which at one point were banned in the US for causing cancerous tumors in lab rats in a rare example of the US government taking the dangers of toxins seriously!

If rapid weight loss still doesn't sound so awful, just wait—it gets worse. We know now that cleanse and detox diets can release chemicals into your body that are generally dangerous to your system. But they are actually linked to a specific deadly condition as well. Doctors affiliated with the University of Southern California report that cleanse diets are linked to what they call a *silent killer*: a condition called *nonalcoholic fatty liver disease* (NAFLD). The liver, as I wrote earlier, is responsible for handling the toxins in our body, a role that makes its health essential. When we give the liver more toxins than it can handle, then it can stop functioning altogether. As if that weren't bad enough, the USC doctors pointed to a Mayo Clinic study showing that NAFLD can mean a higher risk not just of liver cancer but of other cancers as well. This should come as no surprise since so many toxins have passed through the poor liver.

Needless to say, dietary detoxification isn't all it's cracked up to be. While those who advocate it are right about how important it is that we remove toxins from our bodies, they are going about that essential task through channels that are wrong and even dangerous. What we need to is to detoxify ourselves on a metabolic level.

What to Expect from Metabolic Detoxification

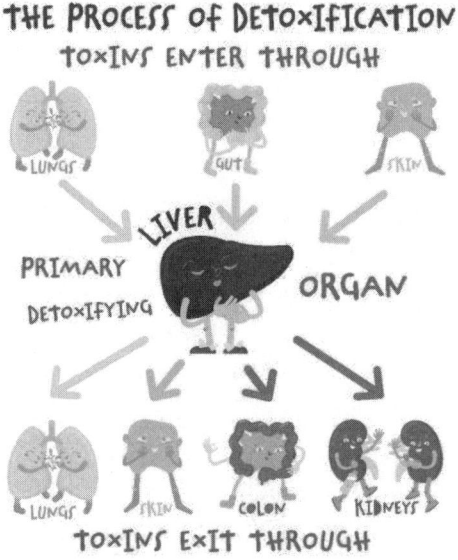

Double Brain / Shutterstock

The idea of detoxification has reached an all-time popular high. Go into any health food store, and you will see powders and potions galore that promise to purge the toxins right out of your body in just seven days. Pick up a magazine while you're on line at the grocery store and read all about a celebrity's miraculous weight loss after attending a detox retreat. While cleanses and fasts are all the rage right now, a person with serious health issues who commits to a metabolic detoxification protocol might not experience that same Hollywood appeal. In fact, similar to the trendy cleanses I just talked about, what happens during the initial phases of detoxification is not sexy at all.

If you decide to commit to metabolic detoxification, the first side effect you most likely will experience is a profound loss of energy since your body will be working harder than it has been to get rid of toxins. You could experience fatigue for two weeks to a month, have altered bowel movements, a decrease in appetite, weight loss, or a slight headache. Your liver and mitochondria will need extra support to perform the detoxification functions they are slowly being allowed to perform. You will feel as if your body is working overtime because it will be. You will need plenty of rest and also a kick-ass multi-supplement to replenish all of the vitamins and minerals that will be pulled out of circulation with the toxins.

You might also experience even more inflammation and swelling in new places at first, as the toxins begin to be released throughout your body. An anti-inflammatory diet, therefore, is key to mitigating some of this extra discomfort. I will go into further detail about diet later in the book. For now, the main foods you will have to avoid are most dairies, white flours, white sugars, and most grains. Eating clean, organically-grown produce and organically-raised meat is also an essential part of bringing down the inflammation.

Because of the expansive scope of care involved in some of my metabolic detoxification protocols, I want to make sure patients are ready to commit their time, energy, and money to their personalized regimen. I include on the new patient intake form the following self-assessment to help determine their readiness. You can take it too, rating each answer on a scale of 5 (very willing) to 1 (not willing)

In order to improve your health, rate how willing you are to:

1. Significantly modify your diet.
2. Take nutritional supplements each day.
3. Keep a record of everything you eat each day.
4. Modify your lifestyle (i.e. work demands, sleep habits).
5. Practice relaxation techniques.

6. Engage in regular exercise.
7. Have periodic lab tests to assess your progress.

Omar, the COO of a large chemical company in Riyadh, called my office one day, hoping to get relief from severe swelling in both feet that started after the removal of his prostate. Apparently, although the doctor who had performed the operation had promised to remove only two or three lymph nodes, the medical report indicated that thirteen lymph nodes had actually been removed. I told Omar that the removal alone of lymph nodes would not cause this amount of swelling. Clearly, he had toxins in his system that were exacerbating this reaction.

Since Omar was a telemedicine patient, I could not perform Lymph-Biologics™ on his feet. After receiving a urine and stool sample from him, however, I saw that he contained a significant amount of heavy metals. I put him on a regimen of metabolic detoxifiers and a strict anti-inflammatory diet. In a few weeks, Omar was able to take off his compression socks for the first time in months and not see little ravioli on the tops of his feet when he awoke each morning.

The Dynamic Duo of Lymph-Biologics and Metabolic Detox

I have patients all over the world whose toxicity levels and lymphedema symptoms have improved dramatically merely from metabolic detoxification protocols. Without the help of Lymph-Biologics™, it just takes longer—six to eighteen months—to see these results. Detoxification is essential not only for lymphedema and cancer patients. Over and over, it has been proven to help patients suffering from chronic diseases and conditions including allergies, anxiety, arthritis, diabetes, headaches, heart disease, high cholesterol, digestive disorders, and obesity, to name a few. Patients suffering from autoimmune disorders have seen astounding results by using detoxification protocols. Undeniably, if you have been diagnosed with fibromyalgia, chronic fatigue syndrome, rheumatoid arthritis, lupus, inflammatory bowel disease, multiple sclerosis, Guillain-Barré syndrome, or psoriasis, you will benefit from a metabolic detoxification regimen.

Detoxing: The Road Less Traveled

Considering the amount of toxicity Americans literally are swimming in, it's no wonder most people find metabolic detoxification protocols challenging. The cost of taking supplements, finding the right water filtration system, using all-natural products, eating organic, and paying out of pocket for an alternative healthcare practitioner with a holistic approach can be unaffordable for many people. Even those who have the financial means to support a healthier lifestyle are not free from the dangers of toxicity. Many know very little about the chemicals they are taking in on a daily basis, from beauty products and home furnishings to restaurant food and even some of the most expensive bottles of wine. Lastly, some people opt out of detoxification because they find themselves too busy to keep up with a healthy lifestyle. They choose to bury their heads in the sand and hope that ignoring what they know will be enough to keep inflammation and health issues at bay.

Recently, Dimitri, a Greek gentleman, came to me because of severe aches he was having in his neck and upper back. He was a busy man with lots of ambition, but he was about one hundred and fifty pounds overweight, and the extra pounds were taking a toll on his body. When I tested his cells, I saw that they were in surprisingly good shape, which I attributed to the healthy Mediterranean diet he'd grown up with. Fresh vegetables, olive oil, and fish are all wonderful foods to start out eating as children and young adults. But now, Dimitri was in his early forties and had strayed from that type of nourishment. Instead, because of his stressful life, he was used to takeout, fast food, and lots of caffeinated beverages.

I told Dimitri that the state of his cells was a gift that he needed to stop abusing with his fast-food diet and stressed-out life style. "One day," I warned him, "that house of cards is going to come tumbling down if you're not more careful." He and his wife were very busy with the new home they were building for their family, and I hoped this analogy would hit home for him. For a few months, Dimitri listened to me, changed his diet, lost some weight, and came for regular adjustments and check-ins. One of the biggest changes he made was he started drinking a lot more water. But sadly, a leopard doesn't change its spots very easily, and soon, he was back to his old bad eating habits and stopped coming for office visits. Like many people, Dimitri, in the end, cared more about the siding on his McMansion than he did about the internal foundations of his own body.

Artist Shalom Gorewitz Says…

> *I am seventy-two years old and have had profession-related physical problems for more than fifty years. During the last twenty years, doctors and physical therapists told me I have arthritis and should learn to live with it. Chiropractors, sport doctors, and acupuncturists have helped with some of the problems, but no one got to the roots until I put myself into the hands of Dr. Loretta Friedman. After about one month of physical manipulation of my disks, several sessions of lymphatic drainage for detoxification, heavy doses of supplements, and improved diet, I have lost more than ten pounds (easily) and generally feel better in places that I had given up on like my knee, shoulders and lower back. I no longer have the symptoms of arthritis.*
>
> *Dr. Loretta's vast technical and practical understanding of the anatomy and structure of the body is matched by the "muscle" in her manipulation and her energetic dance of healing. She touches energy points or brushes her hand with a faith healer's sweep. She finds the places that need mending and is somehow able to nudge the disks back in place without popping and cracking. The lymphatic drainage process is painless. Dr. Loretta claims this is key to unlocking the cells from the toxins that are blocking healing cell activities.*

Do I still have some aches and pains? Yes. Some of them are the good kind from exercising and having more confidence in the body. Some are echoes of past stiffness and discomfort. I think Dr. Loretta would say even she can't cure the accumulation of aching during seventy years of use in a month. In short, I trusted her, put myself in her care, listened to and followed her instructions and wisdom, and worked with her through lifestyle changes and new disciplines. Was it worth it? Yes. This "reset" has been a revelation and blessing.

What You Need to Know

- Instead of treating toxicity, most doctors employ modalities that provide patients with only short-term relief from symptoms.
- Lymphatic massage alone cannot remove the toxins in your body, nor can it open the blocked ducts of your lymphatic system.
- Lymphedema occurs when toxins do not allow the lymph fluid to drain from internal tissue.
- Lymph-Biologics™ is the only proven technique that actually resolves inflammation and swelling of the lymph nodes.
- Lymph-Biologics™ helps rid the body of toxins by increasing fluid volume and exchange at the site of the lymphatic vessels.
- The question to ask isn't *Are you toxic?* but *How toxic are you?*
- Shrinking cell membranes and hindered cell receptors are by-products of high toxicity levels.
- Detoxification methods of healing have been used for thousands of years.
- Metabolic detoxification helps patients suffering from allergies, anxiety, arthritis, diabetes, headaches, heart disease, high cholesterol, digestive disorders, and more.

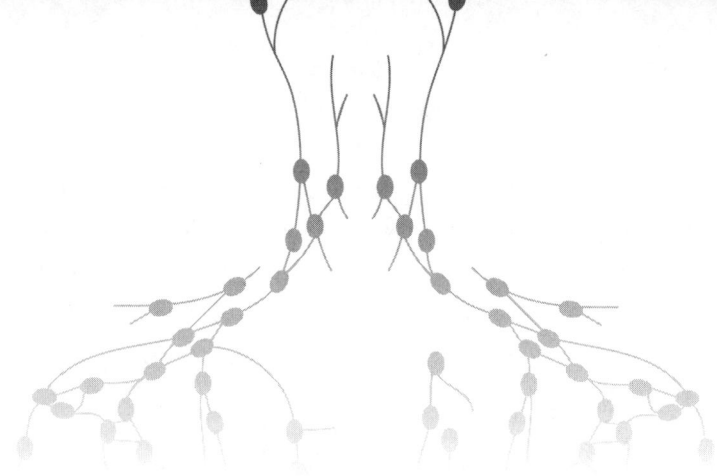

CHAPTER 8

What Your Cells Can Tell You That No Doctor Will

In my practice, I have yet to see a cancer patient who does not have an astronomically high toxicity rate. In fact, I'd say that the correlation between cancer and the amount of toxins in your system is undeniable. Obviously, cancer has existed in humans since before recorded history. But we now live in a world that is full of environmental chemicals having a toxic effect on our bodies—and nobody denies that they are leading to greater cancer rates. In fact, the World Health Organization estimates that as much as 19 percent of cancer cases are directly due to environmental exposures, or, in other words, toxins. One form of cancer that they point out has actually gone down recently—by 25 percent over the last five years—is lung cancer. The reason for this decline is that the rate at which people smoke has declined. Of course, the chemicals from cigarettes that we inhale when we smoke are prime examples of toxins.

A good cell test could be one of the most reliable ways to determine whether or not a disease like cancer will return. It's also why that's one of the first things I do when a patient comes in for an office visit. But even before the cell test, which we will get to soon, I ask my new patients to fill out a twenty-three-page self-evaluation form (found in the back of this book) that, to many, feels like a test onto itself. This form is not like any other medical history form you'll come across. It's more like a magnifying glass that forces you to see the good, bad, and the ugly of your lifestyle choices. The following are just some of the questions you'll find if you decide to fill out this form yourself. As you'll see, I start out easy:

- When was the last time you felt well?
- What makes you feel worse?
- What makes you feel better?

But even the easy questions can feel challenging for patients who have been suffering for a long time and can't remember the last time they felt well. They have also been in so much discomfort that they can't tell what feels better or worse. (Does sitting help? Or is standing better? What brings relief? Ice? Heat? Do those compression garments feel better on or off?)

Then come the questions that might make you blush or squirm a little in your seat:

- What type of contraceptive do you use? Condoms? Diaphragms with spermicide? IUDs?
- How would you describe your stools lately?
- Have you experienced genital pain? Prostate enlargement?
- Have you ever had a problem with alcohol or other recreational drugs?

I ask these types of questions not because I'm nosy but because I need to assess all of the imbalances in a patient's body, whether brought on by hormonal shifts or substance abuse. Many people don't think twice about the birth control they have been using since they were eighteen, which could very well be throwing off their whole system.

Many questions on the questionnaire fall into the "I have no idea" category:

- How many times in your life have you taken antibiotics?
- How many times have you taken oral steroids?
- Do you have mercury fillings in your teeth?

These questions are usually left blank, even though they are important since they have mostly to do with toxicity exposure. It's a frightening fact that the more specific my questions get about patients' exposure to toxins, the more blanks I see on the intake form. People tell me all the time that some of these questions—what I call the "fine-tooth-comb questions"—catch them off guard because they illustrate just how many toxins we are all exposed to each day. As you already learned, the toxins listed on my intake form are just the tip of the iceberg.

Though the questions above may look random to you, I have designed them (and the others in my questionnaire) carefully in each case to pinpoint a source where dangerous toxins could be entering my patients' bodies. Antibiotics, for example, are often looked upon as not only harmless but helpful because they have been prescribed by a physician to treat a health problem. In principle, this outlook is correct. When antibiotics save us from a dangerous infection, they are a miracle of medicine. However, on the other side of the coin, they are still outside chemicals introduced into the system and, as such, have the potential to become toxins and do harm to us. In fact, the New Zealand Medicines and Medical Devices Safety Authority/Medsafe has gone so far as to warn people that antibiotics are "a common cause of drug-induced liver injury." They go on to warn that drug-induced liver injury is "unpredictable and largely dose-independent," which means that there is no "safe" way to take antibiotics that will reliably eliminate the danger—every time you do it, you are taking a calculated risk that the toxins in the drug (which is hopefully simultaneously curing you of some other ailment) may damage your liver.

The threat of dangerous effects on the liver is not confined to rare or extra-strength antibiotics or to only a few people. Antibiotics account for 45 percent of all drug-induced liver injury, and they are, as a class, its single greatest cause. Even an antibiotic as common and frequently prescribed as, for example, amoxicillin prompted a doctor to warn in a recent article for GoodRX Health that "Liver damage from this antibiotic can occur shortly after you start taking it and can be prolonged. Signs of liver injury are often detected even after patients stop the medication."

If, as we have seen, antibiotics can be a double-edged sword with such sharp edges, it should come as no surprise that oral steroids can also act as toxins, packing some unpleasant surprises for our systems. Specifically, anabolic steroids can cause an awful condition called cholestasis, in which bile—a fluid made by the liver, which is used in digestion—is blocked from going where it needs to go, so it ends up in the blood instead. That's literal blood toxicity that can be caused by the steroids that any of us might be prescribed at any time. In addition, Carl Grunfeld, chief of the metabolism and endocrine section of the San Francisco VA Medical Center, did a study finding that within only twelve weeks, anabolic steroids increased not only serious liver toxicity, but also increased LDL ("bad") cholesterol and decreased HDL ("good") cholesterol. He reported, "We found grade III and grade IV liver toxicity in some men, which means a very significant risk of serious liver damage.... The higher the dose, the more the toxicity." For once, it seems, the medical establishment may be right when they warn about the dangers that lurk in children being tempted to use steroids. And yet, as with antibiotics, most people happily take steroids whenever mainstream doctors suggest them without worrying about the danger of toxicity.

By now, I am sure you can guess why I might have asked about mercury filling in teeth. Mercury is a well-known toxin with respect to the human body, yet many people think nothing of the mercury fillings that they carry around every day in their mouths. I find it baffling that, despite the known dangers of mercury exposure, dentists continue to use mercury in amalgam fillings, which consist of half mercury and half silver. You may think that the silver filling would prevent any mercury from escaping, but you would be wrong. The government of Canada released an official report confirming that mercury vapor is continually released from these fillings, that more of the liquid metal can be released through wear or corrosion and swallowed, and that even more can be released and embedded in the body whenever old fillings are removed.

The International Academy of Oral Medicine and Toxicology reports that exposure to this dental mercury has been linked to a long list of awful diseases, including Alzheimer's disease, Lou Gehrig's disease,

Parkinson's disease, multiple sclerosis, and more. While the U.S. Food and Drug Administration still for some reason allows mercury to be used in fillings, they have released a long list of categories of people who are warned against getting them, including pregnant women, women planning to become pregnant, new mothers, and anyone with impaired kidney function, Parkinson's, or "heightened sensitivity" to mercury (although there is no known safe amount of mercury to be exposed to!). If the fillings are truly safe, why aren't they safe for everyone?

If that information about mercury has you scared away from getting a mercury filling (or disturbed by one that is already in your mouth), consider this: a study connected with the United States Geological Survey found that 27 percent of fish from 291 streams in the United States contained more than the recommended limit of mercury! I wasn't kidding when I said that we are surrounded by toxins and constantly exposed to them. Those are just three of the many factors I check for in my exhaustive questionnaire. Taken as a whole, it is designed to give me as complete a picture as possible of just what toxins a patient might have absorbed into their body unknowingly.

Reading Between the Lines

You would think that as a Directional Non-Force Technique (DNFT) chiropractor, my first go-to in alleviating a patient's discomfort would be to use my hands to correct a spinal misalignment or a disc. But for lasting relief, you need to find the source of the symptoms. More often than not, the problem starts from a place beyond what the hands can feel or the eye can see.

The cell test I do in my office is similar to the bioelectrical impedance analysis (BIA) physicians use to assess body composition. That test simply uses electrodes and an electric current to determine how much of a person's body is composed of fat. While fat composition is an important factor in determining a person's health (in addition to toxicity, as we have already seen), I look at considerably more information in the results of my cell test. For one, the test is an important method in determining whether a patient is predisposed to ailments such as cardiovascular disease and diabetes. Most doctors who read the results are looking for the ratio of body fat to lean muscle tissue and go no further. But so much more information is available about a patient's health when we take the time to read between the lines.

There are a few signs I'm looking for when a patient comes into my office with low energy, swelling, inflammation, aches, and pains. Some of the culprits may surprise you.

Eat More Fat (the Good Kind)

The shape of a person's cell membrane, the outer lining of the cell through which fluids and nutrients are constantly passing, should be round, plump, and grape-like. Too often, however, the cell membrane looks more like a shriveled-up raisin or prune.

The good news is that this condition is not permanent and can be remedied by a few simple changes in diet, including stocking up on Omega-3, the good kind of fat. What's so good about Omega-3? It's essential to running all kinds of important reactions in the body, including Cyclooxygenase 2 (COX-2) reaction, which reduces inflammation. It's also a powerful brain booster that can help with memory loss and all types of other cognitive problems like lack of focus and concentration. With a good supplement along with incorporating more wild cold-water fish, flax seeds, and chia seeds into your meals, you can reduce signs of aging on a cellular level. And who wouldn't want that? (More on Omega-3 and how to boost it in Chapter 10.)

We live in a fat-fearing nation that only recently has gotten the memo that some fats are actually beneficial and essential to your health. As is the case with many trends, most people go overboard with this. For example, foods like avocado and coconut oil have become household items that line the pantry shelves of health-conscious millennials. However, what these thirty-somethings don't know is that these foods, while helpful in some ways, don't necessarily promote cell health and can be hard on the liver, pancreas, and gallbladder when consumed in large amounts. Avocados contain two chemicals called estragole and anethol, which can cause liver damage, as well as carbohydrates called polyols or sorbitol, which can affect people with sensitive stomachs or irritable bowel syndrome. As a result of this information gap, many young people who come to me are overdosing on these superfoods, unknowingly creating chronic musculoskeletal problems.

You may be wondering how eating a bowl of guacamole could possibly cause back pain. Here's how. We know that the brain controls the body by sending messages down the spinal cord and out to different areas. The Metric Chart, first published in 1963, shows just how specific these messages can be. In fact, according to the chart, each of our internal organs is affected by a specific vertebra of the spine, and vice versa. When the body has to work overtime to digest certain fats, these fats end up co-opting organs such as the pancreas and gallbladder, causing inflammation and muscle strain around the specific vertebrae that support them. The resulting subluxations, or misalignments, can be excruciatingly painful.

Jennifer, a young woman from New Jersey, came to see me, complaining of chronic neck and shoulder pain. I'd seen this many times before: women and men who seemed strong and fit whose musculoskeletal structure was off for no apparent reason. Before I made any adjustments to Jennifer using Directional Non-Force Technique, I asked her about her diet. She admitted that she existed mostly on processed crap from her office cafeteria during the day and takeout dinners at night. Because of her poor diet, Jennifer had developed digestive issues that she was treating with Nexium, one of the latest proton pump inhibitors put out by the drug companies. Prilosec and Zantac are other versions of this drug used by millions of Americans to mitigate acid reflux and heartburn. However, like most medications designed to alleviate symptoms, it was bypassing the cause.

I told Jennifer to toss out the Purple Pill and prescribed a regimen of potent digestive enzymes that mimic the elements that should be found naturally within the body. I also told her to start eating more whole foods, including those that contain natural anti-inflammatory properties.

(More on these in Chapter 10.) Within a week, not only was her neck and shoulder pain significantly reduced, but her acid reflux was ancient history. After only six visits, Jennifer was back at the gym, doing light workouts again—only now, she was pain-free.

The Stress Snowball

Now let's talk about stress, another inflammatory element that can wreak havoc on the body. Stress kills. I mean it. A marathon runner walks out of his doctor's office with healthy EKG results in his hand and in the parking lot drops dead from a heart attack. We've all heard shocking stories like this one about people who appeared healthy and fit, but who suddenly died without warning. I have no doubt that the underlying cause of many of these cases is the endless amount of stress, whether emotional, physical, or environmental, our bodies go through each and every day.

One telltale sign of chronic stress is the constant fatigue and malaise that plagues people as young as those in their twenties. Maybe you can personally relate to or know someone who in only a matter of a few years went from the Energizer bunny to a certified couch potato. This person used to go out dancing after work, jog in the park five mornings a week, and volunteer at a local charity on weekends. Now, she has just enough energy to drag her butt home from the office, order Chinese food, and fall asleep in front of the television. This person isn't lazy. Nor is she old. She isn't even overweight. She is suffering from a pandemic that's run rampant throughout our country since long before COVID-19 showed up. This person is a victim of the hidden pandemic of stress.

In order to understand just how stress affects our energy level, let's take a look at how our bodies produce energy on a microscopic level. The mitochondria are the powerhouses of the cell. It's where the Krebs cycle (also known as the tricarboxylic acid cicle) takes place, and where adenosine triphosphate (ATP), an important source of energy, is produced. Ideally, this ATP gets restored at night, while we are sleeping. However, many people are too stressed to get enough sleep to replenish it at sufficient levels.

Barbara was a patient of mine in her early fifties who had been commuting from Staten Island to her job at Price Waterhouse in Manhattan for nearly thirty years. The stress of her daily schlep into the city and her high-profile position in human resources was depleting her of all her energy. A quick cell test confirmed my suspicion. To put it bluntly, Barbara's mitochondria were in the toilet. I put her on a regimen of ubiquinone and a super antioxidant powder, and, within a week, Barbara reported back amazed how energy had skyrocketed.

Barbara was so pleased with how good she felt that she invited me to speak to the national HR directors at Price Waterhouse, a team consisting of about one hundred men and women. I was a co-presenter with my friend Sharon Melnick, a well-known psychologist and executive coach. Melnick is something of an expert on stress herself, as the author of *Success Under Stress: Powerful Tools for Staying Calm, Confident, and Productive When the Pressure's On*. She's known for using her professional expertise to help people decrease the stress in their lives to healthy levels. As she puts it, "We only experience 'stress' when there are aspects of situations that feel out of our control. The more you can control, the less stress you will experience." I spoke with this group about good eating habits and anti-aging practices. In turn, I was asked to conduct cell tests on a pod of employees who recently had committed themselves in a group effort to improving their health and fitness.

The results of the cell tests reflected astronomically high levels of the stress hormone cortisol. Now, cortisol is a great hormone. It's what kicks your butt out of bed in the morning and gets your day going, reaching its peak level between the hours of 5 and 7 a.m. When a person is under too much stress, as this group of employees clearly was, cortisol levels intermittently rise throughout the day. These higher levels increase your body's production of insulin while simultaneously decreasing your cell-receptor sensitivity and throwing your body's physiological responses completely out of whack. As a result, your glucose and insulin levels elevate, exhausting your pancreas, the organ that produces insulin. These elevated insulin levels also cause your body to store fat that would normally be used for energy. Looking around the auditorium, I saw a sea of sluggish people in their forties and fifties, almost all with an extra twenty to thirty pounds—think muffin top—around their guts. They were the perfect candidates for Type 2 diabetes, a disease that plagues over thirty million Americans.

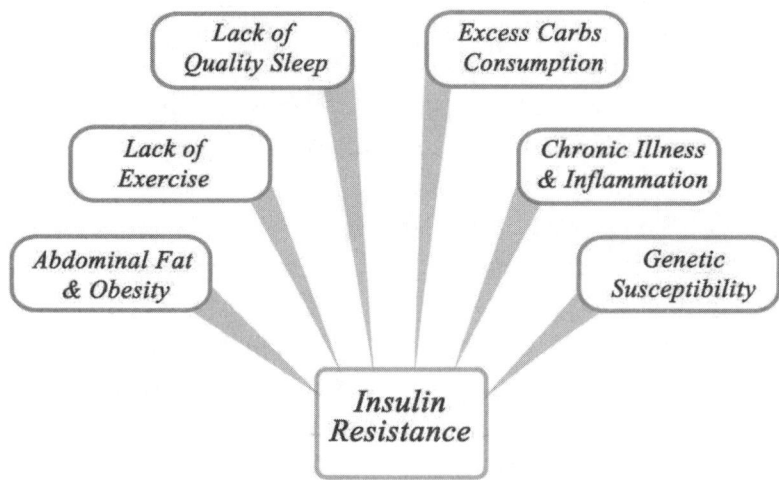

The Skinny on Cholesterol

Imagine a hot blowtorch shooting through your blood vessels. In essence, this is what's happening when your body is stressed. The elevated cortisol and insulin levels that circulate throughout your body all day generate a damaging amount of inflammation in the intima lining of the blood vessels, particularly the tiny ones around the heart. The result is similar to the cracked paint on the interior walls of a house after a fire. In response to the damage, the liver sends out a work crew in the form of cholesterol molecules to provide a new coat of protection. This repair job eventually causes the blood vessels to narrow, restricting blood flow and instigating atherosclerosis, the life-threatening condition otherwise known as coronary artery disease.

The question is what are healthcare physicians doing about this condition that affects more than three million Americans each year? The answer is simple: not enough. Performing routine EKGs and echocardiograms only reveals how well the heart is pumping. Sad to say, clinicians almost always completely overlook the real underlying cause—that is, the condition of the coronary arteries. A good score on a nuclear cardiac test or calcium cardiac (medical jargon for a special kind of x-ray that shows a picture of the calcium-containing plaque in the heart's arteries) is key in detecting potential coronary failure. All too often, however, these tests are done only after a patient has already suffered a heart attack or blood clot. In fact, only a handful of physicians, such as Dr. James Blake, a cardiovascular medicine specialist in New York City, make this kind of testing a regular part of their practice. What's worse, standard health insurance plans won't cover these screenings, making them cost prohibitive to most patients.

Instead, pharmaceutical companies are pushing the sale of statins, drugs that inhibit the liver's production of cholesterol, and doctors have been drinking the Kool-Aid for years. For the past decade, the number for what's considered an acceptable cholesterol level has been decreasing, and statins have been jumping off pharmacy shelves, generating millions of dollars for companies that create them. The use of these statins creates myriad side effects. For one, it depletes the mitochondria of ubiquinone, resulting in fatigue and inexplicable aches and pains.

Like every natural chemical your body produces, cholesterol is there to serve a very important function: namely, to protect us from viruses. Think about the lint at the bottom of your pocketbook, the white stuff that floats around the air above a flock of pigeons on the street, or the scum along the sides of your child's fish tank. All of these elements contain viruses that we are exposed to on a daily basis. The overuse of statins is killing our bodies' natural ability to fight these viruses, often to a lethal degree. For this reason, I suspect that a great number of COVID-19-related deaths occurred in people who had cholesterol inhibitors flooding through their systems, compromising their immune systems and keeping them from naturally fighting the virus.

Every year, more and more strange cases of seemingly-unstoppable viruses and parasites are rearing their heads. I received a phone call from a woman in her forties who had been suffering for years from a strange itching and burning feeling just under her skin. She'd visited every specialist on the block, but no one could help her find the cause of her discomfort. I suggested that she may have a parasite or a virus. A lightning bulb went off in the woman's head and prompted her to do her own research online. She discovered that she had a rare disorder called Morgellons disease (MD), characterized by the presence of worm-like fibers that erupt from unbroken skin and slow-healing sores.

Morgellons disease disproportionately affects women and tends to bring with it a whole slew of problems. Among sufferers, about 70 percent also had chronic fatigue syndrome, 60 percent had ongoing aches and pains, and 60 percent had problems with cognitive functioning. But the people who suffer from this maddening condition are often denied help from the medical community who call the disease "unexplained dermopathy" and, unable to find an explanation, presume that the sufferers are mistaking fabric fibers for symptoms. But it's worth noting that 80 percent of sufferers reported exposure to solvents—in other words, potential toxins. As Mark L. Eberhard, director of the division of parasitic diseases and malaria at the Centers for Disease Control and Infection, put it, "These people are definitely suffering from something. It has impacted their lives greatly."

The woman who saw me for a consultation was so desperate to find a cure that she made up her own solution of diluted bleach and poured it on her arm. The frightening results were nothing short of a horror movie. Tiny particles that looked like bugs started emerging from her skin. The woman called me the next day, thanking me for my help in getting to the bottom of things. I told her that she'd do well to get on supplements to boost her immune system and lower her body's inflammatory response to keep any other microscopic intruders at bay.

Another important fact about cholesterol that most people don't know about is its role in the synthesis of sex hormones. It's no coincidence that a man who has been taking statins to lower his cholesterol is also exhibiting lower-than-normal testosterone levels. Big Pharma's answer to the resulting cases of erectile dysfunction that one-third of American men experience, of course, is popping another little pill. But Viagra, a.k.a. "Vitamin V" or the Blue Pill, is by no means a cure. The ironic ending to this saga of prescriptions is a whole populace of middle-aged men walking around with nothing to do with their twenty-four-hour erections because they still suffer from chronically-low sex drives. (Just because they can get it up doesn't mean they have the desire to stick it anywhere.)

The Magic of Ratios

It's possible to be a size two and still have the body fat-lean muscle-tissue ratio of an obese person. These are the people I call the Fat Skinny People of the World. Take the Ladies who Lunch crowd, the ones who take spin and boot camp classes several mornings a week and then meet their friends for lunch, which consists of a few pieces of lettuce, a four-ounce piece of fish, and a couple of glasses of wine. Even if the number of calories they are consuming is sufficient, they are starving themselves of the protein they need to support their workouts. Their bodies still look great in a tank top and pair of spandex leggings, but their muscle tone is far from what it should be considering the amount of time spent at the gym. Peel off the leggings, and you'll find the cellulite and loose skin.

Another important ratio I look at in cell test results is a patient's complete body fluid (i.e., from organs, connective tissue, and lymphatic system) versus the fluid that is floating around outside of these cells. The ratio should be as close to 1:1 as possible. If more fluid exists on the outside, then, more than likely, some type of toxicity is present. The heavy metals that leach out of the ground on a construction site or from a car exhaust that gets released into the air are enough to contaminate our bodies. Mercury, arsenic, cadmium, lead, and chlorine are only a few on this list that can and should be treated by metabolic detoxification supplements and chelators that bind with the toxins and usher them out through the urine and the feces.

Factors such as the environment, chemical sensitivities, medications, and a person's natural predispositions all play into how affected by these metals a patient can be. They also determine whether or not metal toxicity will occur again in an individual. The first step to purifying our systems is to find out just how toxic we are by, again, conducting that all-important cell test. Once toxicity levels are established, urine tests can determine a toxic metal profile to give you a better idea of the kind of toxicity you have. You can also conduct a stool test to detect leaky gut syndrome.

If there are toxins present in your body, you can be certain they are compromising your lymphatic system, causing blockages and stagnation that negatively impact your immune response. In the next chapter, we will dive into why Lymph-Biologics™ is one of the most effective methods for getting rid of toxins and moving lymph fluids. I'll also give you specific exercises you can do at home to start getting relief from lymphedema on your own.

What Joy Says...

I'm sixty-nine, and I used to pride myself on looking and feeling ten years younger. But then I began suffering from what I thought were digestive issues. I had my appendix and gall bladder removed, and after that my lower back pain became almost unbearable. At one point, I couldn't even bend my torso enough to get out of bed. All night long, I was waking up with pain throughout my legs and arms. My joints were so swollen that when I wrapped my hand around my wrist, my thumb and forefinger couldn't touch.

I met Dr. Loretta while I was on vacation in Italy and told her about my symptoms. Until then, I thought that everyone lived with pain the way I did. I became a telemedicine patient of Dr. Loretta's, and she explained to me that my digestive system had been working so hard that it was causing my whole body to become inflamed. She prescribed just two supplements to take regularly each day, one formula of bioactive pancreatic enzymes and another that contained lipotropic nutrients. After just three days, I felt so much better! My quality of life returned. I no longer needed a dog walker to walk my giant Weimaraner. I could walk my dog myself! And now every morning, I roll out of bed, like a normal person. For me, age is arbitrary once again.

What You Need to Know

- There is a direct correlation between cancer and the amount of toxins in your system.
- Natural remedies such as Essiac tea and sessions inside an O_3-infused "sweat box" help release the carcinogenic toxins out the body's pores.
- The cell test I perform on patients is similar to the bioelectrical impedance analysis (BIA) physicians use to assess body composition.
- Omega-3 is essential to running important reactions in the body, including cognitive functions and COX-2 reaction, which reduces inflammation.
- "Superfoods" such as avocado and coconut oil can be hard on the pancreas and gallbladder when consumed in large amounts and can contribute to back pain and digestive issues.
- Stress increases cortisol levels and throws your body's physiological responses completely out of whack.
- Stress elevates glucose and insulin levels, exhausting the pancreas, the organ that produces insulin.
- Elevated insulin levels also cause weight gain, a telltale sign that a person is stressed.
- Raised cholesterol levels cause the blood vessels to narrow, restricting blood flow and instigating coronary artery disease.
- The Fat Skinny People of the World are starving themselves of the protein needed to support a healthy muscle-to-fat ratio.
- A cell test will help determine environmental toxins, chemical sensitivities, medications, and a person's natural predispositions, factors that all play a role in a person's toxicity level.

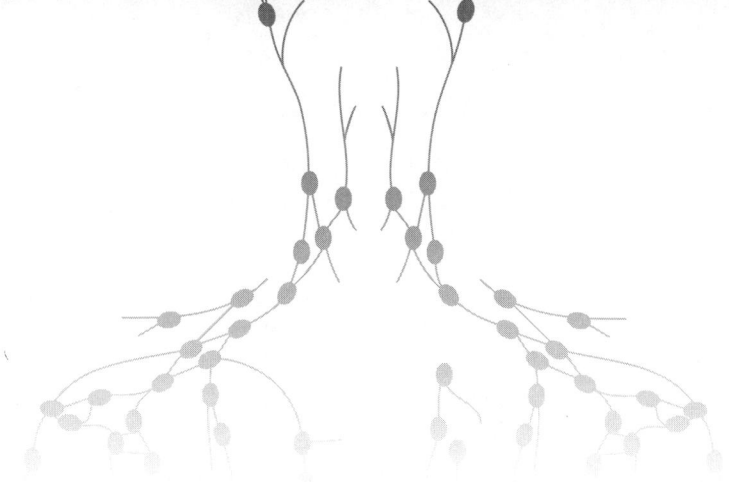

CHAPTER 9

It Takes Two: Your Breasts and the Lymph Link

Your Breasts Are Talking to You

Hopefully, by now, your eyes have been opened to the underlying toxins that infiltrate our daily lives. If you are a woman reading this book, I'd like to open your eyes even further to the important role your breast health plays in your body's overall wellbeing. Contrary to what western medicine tells us, our breasts are more than just two lumps sitting on our chests with the potential of making us sick. They are only nature's most well-designed manufacturer of all the life-giving nutrients that humankind needs to survive. Think about it: a mother's breast milk contains the perfect mix of vitamins, protein, and fat that a baby needs to grow as well as all the antibodies that help fight off viruses and bacteria. On top of this, since breasts contain a high density of fatty tissue and lymphoid, they have a high concentration of ducts and glands that act as the greatest built-in storehouse for all this good stuff. This property makes your breasts your body's greatest ally—but also potentially its worst enemy when it comes to toxicity.

The fact is, your breasts are where your body likes to hold onto a lot of *schmutz*. That's the scientific word for environmental pollutants and DNA-damaging toxins. In particular, certain pesticides and heavy metals tend to gravitate toward the fatty tissue in breasts, as proven by numerous studies that show high levels of such toxins in breast tissue and breast milk. In just one recent example, a study published in the *Environmental Science & Technology Journal* checked women's breast milk for the presence of toxic chemicals known as PFAS (per- and polyfluoroalkyl substances). These are a group of about 9,000 substances known as *forever chemicals* because they accumulate in humans and do not naturally break down, so they stay in your body "forever." They appear in extremely common sources that we all come into contact with such as food packaging, clothing, carpeting, water, and stain resistant chemicals, and they have been linked to cancer, liver disease, thyroid disease, and other dangerous conditions. The study detected the PFAS in every single sample tested—and at a level almost 2,000 times higher than what is considered "safe" in drinking water.

But PFAS are not the only group of chemicals increasing our toxicity levels. As you learned about in an earlier chapter, endocrine disruptors form another group of chemicals that pose an issue for breast health and other hormone-related areas, such as prostate health in men. These chemicals, found in plastics and food packaging, mimic estrogen and other hormones and get distributed throughout your body or simply accumulate in your breasts. They are what's creating hormonal imbalances in men and women and creating issues with fertility. BPAs can cause too much estrogen in men and premenopausal women—a condition you already learned about called *estrogen dominance*. It causes toxic fat gain, water retention, bloating, and a host of other health and wellness issues.

For older women, the story is grimmer. As women age, a natural decline in testosterone and progesterone levels leaves a relative excess of estrogen, so a hormone imbalance could be one of the leading causes of breast cancer. I wrote this book to tell you the things most doctors won't tell you. One of my biggest messages is this: *women need to start paying attention to their breasts*. I'll say it again: *pay attention to your breasts*. Not just once every few years when it's time for a mammogram, or when it's time for a breast exam. Most women's breasts actually talk to them once a month when they start to get achy and swollen during their menstrual cycles. Many girls grow up thinking this is a normal part of their cycle. I am here to tell you that, despite what your doctor and Google search results tell you, it is not. In fact, lumpy, bumpy, tender breasts are a clear sign of toxicity and hormone imbalance that should not be ignored.

In case you missed it in high school, here's a crash course in what is supposed to happen to your hormones during a normal monthly menstruation cycle. At the end of your cycle, once your body is finished shedding the lining of the uterus during the stage we call our *period*, the uterine lining starts to build up again. Slowly, your hormones return to their normal balance. Your progesterone levels increase, as your estrogen levels decrease. However, more often than not, this natural lowering of estrogen does not occur in women. Instead, most women become what's known as *estrogen dominant*, a condition that causes irritability, painful, tender breasts, and an array of other unpleasant symptoms.

Now, just because a condition is common does not mean it is normal. Modern society uses a lot of smoke and mirrors to make all kinds of things appear normal. As we already know from the fake diagnoses I mentioned in a previous chapter, Big Pharma loves slapping a nice, pretty label on a group of symptoms and then coming up with a dozen or so remedies for it. Premenstrual Syndrome (PMS) is another example of these fake diagnoses. It normalizes the uncomfortable bloating, headaches, cramps, and mood swings that accompany seemingly healthy women and has become a rite of passage for adolescent girls in this country—not to mention an excuse to get out of gym class. What is the medical industry's answer to these symptoms? A regular dosage of ibuprofen or naproxen. As you will recall, I have already discussed the potential dangers of toxicity that these pain killers bring with them.

People talk about cramps. They talk about bloating and mood swings. But no OB/GYN I know routinely discusses estrogen dominance. What causes this common hormonal imbalance in most western women? Typically, a combination of two or more of the following: chronic stress, poor gut and liver health, and the environmental toxins known as xenoestrogens—foreign substances that enter the body and act as estrogens because they are close enough to natural estrogen in molecular structure.

They can come from common sources such as plastics, chemicals, pesticides, and even our water systems. And when they load our bodies with more estrogen than usual, they become toxins. How do you know if you could be estrogen dominant? Here is a checklist of some common symptoms in estrogen-dominant women:

- Bloating
- Swelling/tenderness of breasts
- Decreased sex drive
- Lumpy breasts
- Hair loss
- Weight gain
- Mood swings
- Memory problems

Sound familiar? Especially at a certain time of the month? Before you reach for that Midol, you may want to check your diet, stress, and toxicity levels.

Later in the chapter, I'll cover breast health in detail, but for now, here are four steps you can take to correct estrogen dominance, a leading cause of breast cancer:

1. Take Care of Your Liver

Since the liver breaks down estrogen, alcohol, and drugs, any other factor that impairs healthy liver function encourages an estrogen build-up.

2. Increase Healthy Bacteria

Keeping bacterial balance in the gut is important to ensure the proper elimination of estrogen from the body via the digestive tract. Try including a daily probiotic in your diet.

3. Boost Your Fiber Intake

Insoluble fiber binds to excess estrogen in the digestive tract, which is then excreted by the body.

A fiber supplement can also affect the composition of intestinal bacteria and reduce the build-up and re-absorption of free-floating estrogen.

Good sources of fiber include brans, the skins of organic fruits and vegetables, nuts (especially almonds), seeds (particularly sunflower seeds), dried beans, and whole-grain foods.

4. Go Organic and Hormone-Free

Most dairy and meat products these days contain added hormones, so choose organic dairy and meat that's labeled *no hormones*.

Why Mammograms Can't Save Lives

The national cost of breast cancer was 16.5 billion dollars in 2010, according to numbers from the Centers for Disease Control. And in 2020, the estimated cost rose to a staggering 20 billion dollars. That means that breast cancer has the highest treatment costs of any form of cancer in the United States. This makes sense, since there are 255,000 new cases in women and 2,300 new cases in men each year, according to the CDC. Over 300,000 new cases of invasive breast cancer will be diagnosed by the end of this year. Leading causes of breast cancer include the average American diet, inflammatory conditions, radiation exposure, ongoing stress, and especially environmental endocrinal disruptors, such as fossil fuel byproducts, toxic chemicals, and environmental pollutants.

Mammograms have long been the gold standard of breast-cancer detection, with their low-dose x-rays that are meant to detect changes in breast tissue, ideally before they reach malignant status. Since x-rays do not easily penetrate dense tissue, the mammogram machine uses two plates that compress and flatten the breast to spread the tissue apart. If you've ever had a mammogram, you know how unpleasant or even painful this procedure can be when you are in perfectly good health.

However, when you are already experiencing soreness or swellings due to abnormalities in your breast tissue, a mammogram machine can feel like a torture device.

While it's true that regular mammograms have saved women's lives, they are not nearly as effective as you, or most women, may think. In a survey of U.S. women's perceptions of mammography conducted by the American College of Clinical Thermology, 717 of 1,003 women (71.5 percent) said they believed that mammography reduced the risk of breast cancer deaths by at least half. In the same survey, 723 women (72.1 percent) thought that at least eighty deaths would be prevented per 1,000 women who were invited for screening. The truth is, based on US mortality statistics, screening mammography prevents approximately one death per 1,000 women screened.

If you are a woman with dense, fibrous breasts, regular mammography might do very little to protect you. Some experts even argue that mammograms are responsible for more harm than good when it comes to women's breast health. According to one article written by Chris Kresser, an expert in functional medicine and ancestral health, mammograms have the potential for risk. In the recent piece, he writes:

> *Despite this massive increase in the use of mammography, there is a substantial body of research indicating that the widespread, overenthusiastic practice of mammography over the past few decades has had little to no effect on breast cancer mortality rates. In fact, the research indicates that mammography screening may do more harm than good. Mammography has demonstrated a number of adverse effects, including breast cancer overdiagnosis, unnecessary breast cancer treatment, undue psychological stress, excessive radiation exposure, and a serious risk of tumor rupture and spread of cancerous cells.*

In the same article, Kresser cites a thirty-year Danish study that indicated no decrease in advanced breast cancer rates among women who received regular mammography exams. At the same time, he points out that thousands of healthy women are falsely diagnosed with breast cancer each year due to the inaccuracy of mammogram screening results. Imagine the needless psychological stress that these women and their families endure because of a bad diagnosis. And this is true not just for American women. According to a systematic review, one in every three breast cancer diagnoses in the United Kingdom, Canada, Australia, Sweden, and Norway is wrong. Many of the small tumors that mammograms detect would eventually disappear and not turn into cancer if they were left alone.

As Kresser mentions, in addition to overdiagnosis, the actual procedure of the mammogram has long been criticized by experts to have negative and, possibly, lethal effects on those who already have breast cancer. Only twenty-two pounds of pressure are needed to rupture the encapsulation of a cancerous tumor. Normal mammogram equipment applies forty-two pounds of pressure to the breasts, making it highly likely that any existing tumors will rupture and spread malignant cells into the patient's bloodstream. If the needless stress of false positives were not enough to get you to look for alternative methods of cancer detection, this last issue alone should do the trick.

Thermography: The Better Choice

It's no surprise that most women know very little, if anything at all, about the drawbacks and dangers of mammography, considering how many medical protocols are promoted as routine and perfectly safe by physicians and drug companies. Patients' blind faith in doctors' advice and recommendations never cease to amaze me. But we cannot afford to remain ignorant when it comes to breast health. Women need to know that there are choices out there for them. Thermography is one of these choices. Using infrared technology, a good thermogram can detect inflammatory patterns without squashing the breasts or emitting radiation the way mammograms do. Thermograms are so noninvasive and harmless, they can even safely be done on pregnant women or women with implants.

If we are going to move from breast disease to breast health, then early intervention is important. Yet women with dense breast tissue are often neglected altogether when it comes to breast care, because they cannot be effectively screened with mammography. Younger women, too, are overlooked. With over 12,150 cases of breast cancer in women under forty each year in the US, according to the National Cancer Institute, we cannot afford to ignore this age category. In fact, when younger women develop cancer, often the disease progresses much more aggressively and is less likely to respond to treatment. Yet there is currently no routine screening for women under forty. Doctor-prescribed thermography would cover this gap in both categories of underserved women.

Breast thermography looks for heat patterns in the breast that may indicate tumor growth. This type of growth is a physiological process that creates increased vascular patterns that can be detected on a modern infrared camera. Thermographers can identify areas of hot and cold in relation to the opposite side of the body. Areas with decreased blood flow will have colder temperature readings, and areas of increased blood flow will have warmer temperature readings. Cancer cells develop a blood supply through a process called angiogenesis. This increased blood supply produces measurable amounts of heat that can be detected by thermometric devices.

A study at the University of Wisconsin showed that with thermal imaging, 70 percent of tumors are found up to ten years prior to their identification on a mammogram. This study was paralleled in a study performed in the USSR showing the ability of thermograms to detect tumor development in the preclinical phase. Angiogenesis is the key factor in early detection, even before an actual tumor can be seen. This blood supply often increases over time as a tumor grows and is an important part of baseline studies and the ability for them to identify a tumor at the earliest possible stage.

Dr. Patricia Bowden-Luccardi, author of *Thermography: The Fibrocystic and Dense Breast* and a prominent motivational speaker on thermography and breast health, has devoted her life's work to educating women about the benefits of thermography and ongoing breast care. She recommends that women start annual screenings at as young as twenty years old to establish a baseline and know the state of their breasts when the patients are young and healthy. These images are captured in real time from the infrared imaging camera, then stored. All images are kept for comparison with future images so that a baseline can be achieved.

Like me, Bowden-Luccardi hopes to see our country move from a nation of breast disease to breast health. She writes, "Unfortunately, the healthiest habits and the most vital natural substances for sustaining our bodies' vitality are largely marginalized in our society by modern medicine. In fact, many of the most valuable things that can be done to sustain one's health are virtually unknown to the average person." She goes on to describe the preventative value of specific lifestyle and diet choices to decrease breast density, balance hormonal metabolism, and remove environmental estrogens. I will share each of these shortly. But first, a recap on thermograms.

The Benefits of Thermograms in a Nutshell

- Less Invasive—Often, when a suspicious image appears on a mammogram, doctors will perform a biopsy for further testing. A thermography scan, however, provides an additional view of the suspected area for further evaluation. This often removes the need for the more invasive approach.
- No Radiation—Mammograms expose your body to high levels of radiation. Some experts even speculate radiation that is 1,000 times greater than a chest x-ray. Since thermographic screenings are heat based—there is zero cancer risk from radiation.
- Detects Cancer Sooner—Thermograms can identify cancer sooner with more accurate results. Before a tumor develops, new blood cells start to form in the area. These can be seen on a thermogram early on.

- Works on Dense Breasts—About 40 percent of women have dense breast tissue, making it very difficult to get an accurate reading on a mammogram. Since thermograms work off heat, not images, the reading can be done accurately regardless of the density of tissue.
- Pain-Free Scans—If the idea of getting your breasts smashed like a pancake between glass makes you squirm, you'll be happy to know that thermography doesn't do that.
- Safe for All Women—If you are a woman under the age of forty, pregnant, or nursing, exposure to radiation is dangerous. Thermography is a safer choice.
- More Accurate—Often, mammograms detect things that are not cancer. These false diagnoses lead to women getting treatments, radiation, and surgery that is not necessary.

What to Expect from a Thermogram

A thermography scan lasts around fifteen minutes and, unlike mammography exams, it doesn't place any sort of stress on the breast. It uses digital infrared imaging to discover symmetrical changes in the breast tissue. Testing is performed in the technician's office, where the patient will be asked to fill out a breast history form. The patient will be left for ten to fifteen minutes to let her body reach equilibrium with the room's temperature before being positioned in front of the camera (thermography system) to image the upper chest, underarms, and breasts.

These images are captured in real time from the infrared imaging camera, then stored. No images are discarded so that they can later be compared with future images, allowing a baseline to be achieved. The images provide a clear view to vascular patterns, temperature differentials, and possible pathological conditions. Once the images are captured, they will be interpreted by a Certified Clinical Thermographer, who will process and grade the images digitally. After analyzing the images, they are graded using a standardized reading protocol from the Professional Association of Clinical Thermographers (PACT).

What You Can Do Right Now for Your Breasts and Lymphatic System

If there's any message you get from this chapter, it's that no matter how old you are, how healthy you feel, or whether or not you've ever had a mammogram, you need to start paying attention to your breasts. Contrary to how the medical profession views them, your breasts are not just two objects that sit on your chest that can one day make you sick. Most of western medicine is about curing diseases rather than maintaining good health. Besides routine self-exams and mammograms, doctors offer almost no advice to promote healthy breasts.

In addition to the general ways you can lower your body's toxicity, which I covered in Chapter 6, such as drinking lots of filtered water, eating organic foods, and exercising regularly, you can specifically protect your breasts and lymphatic system by taking the following steps highlighted in Bowden-Luccardi's book.

Wear less constricting bras—or, better, no bra at all!

Are you constricting your lymph system by wearing tight bras that contain padding or underwire? Your breasts, like your arms and legs, are appendages, meaning parts of your body that are designed to hang off the main trunk. They were not meant to be squeezed into synthetic material with metal scaffolding that binds them to your body. The way western society treats them like decorations that can be pushed together and propped up just so they are pleasing to the eyes of others is not what Mother Nature planned, to say the least. Since they are technically appendages, your breasts are actually meant to remain a few degrees cooler than the rest of your body. Yet, most women's breasts are pressed up against their heart area and at a constant warmer temperature than they should be.

Patricia Bowden-Luccardi writes in her book on thermography, "Bras are putting our breasts at risk for bad breast health and cause toxins to be concentrated in the breast tissue. Underwires stop just under the armpit where a cluster of lymph nodes reside and block the flow of lymph." Wearing a bra to bed is the most dangerous thing you can do to your breasts since you are giving them no relief from constriction at all. How do you know if your bra is too tight? Red groove marks on your skin should give you a good idea. If your breasts change size throughout the month, Bowden-Luccardi recommends purchasing bras in different sizes to accommodate the fluctuation.

2. Wear loose clothing with natural fibers

Petrochemicals found in synthetic clothing are absorbed through the skin and taken up by the lymphatic system, adding to the toxicity in your body. On top of this, tight-fitting clothing, like bras, only restricts the flow of lymph. Choose natural fibers, instead, such as flax, linen, cotton, silk, and wool.

It sounds incredible, but toxic chemicals can really be absorbed through your skin and cause you serious harm. One study conducted by Stockholm University examined sixty garments that had been sold in shops and found thousands of chemicals in them. Among those chosen for further study because of their potential for toxicity, quinolones and aromatic amines were found. Even organic cotton is not free of risk, but actually contains high concentrations of benzothiazoles. Giovanna Luongo, PhD in analytical chemistry at Stockholm University, concluded: "Exposure to these chemicals increases the risk of allergic dermatitis, but more severe health effect for humans as well as the environment could possibly be related to these chemicals. Some of them are suspected or proven carcinogens and some have aquatic toxicity." Are you getting this? Just as important as paying attention to what goes in your body is what goes on it.

In an article called "Toxic Threads: The Big Fashion Stitch-Up," Greenpeace International confirmed that these hazardous chemicals could be found not just in suspicious, cheap, knockoff clothes, but even in the garments produced by such major and trusted brands. Some of the biggest culprits are any clothes made of acrylic; they contain dimethyl formaldehyde, which the CDC says can interact directly with the skin to cause liver damage and other adverse health effects.

And you may want to watch out for any garment that claims to be static resistant, stain resistant, flame retardant, or wrinkle-free. Those effects can't be achieved without the use of some pretty serious chemicals. According to The IFD Council, the world's leading modest fashion and design council representing the Islamic economy, these chemicals have caused reproductive and developmental defects in rodents. And some of these clothes contain triclosan, which may cause cancer risk after regular exposure, according to data on factory workers.

3. Dry brushing

Since the lymphatic system is very superficial, located right under the surface of your skin, a lot of pressure does nothing to stimulate the lymphatic fluid. In fact, too much pressure on it, say, with a Swedish massage, literally collapses the entire system. Instead, dry brushing should be a light application of fine bristles that stimulate the lymphatic system at the perfect amount of pressure. The principle here is a simple one. Not only does the process stimulate the nervous system, but, as Dr. Shilpi Khetarpal told the Cleveland Clinic, "Dry brushing unclogs pores in the exfoliation process. It also helps detoxify your skin by increasing blood circulation and promoting lymph flow/drainage." Not bad for a practice that takes only a couple of minutes out of your day.

4. Alternate between cold and hot water in your shower

Lymphatic vessels contract in cold temperatures and dilate in response to heat. An alternating cold and hot shower is a kind of hydrotherapy that uses water temperature and pressure to move stagnant lymphatic fluid, increase circulation, and boost immune function and metabolism. Be sure to always end on cold water. Avoid this practice if you are pregnant or have a heart or blood pressure condition.

This approach is scientifically known as contrast bath therapy, and even mainstream science has recognized its positive effects. In fact, studies published in *PLOS One* (a peer-reviewed journal published by the Public Library of Science) and the *Journal of Applied Physiology* both independently showed that showering with alternating cold and hot water positively influenced the immune system, and one showed that the therapy "resulted in significantly greater improvements in muscle soreness at the five follow-up time points."

5. Move your body

Since the lymphatic system does not have a pump like the circulatory system, it is completely dependent on physical exercise to move. Stagnation from sitting all day is a pervasive issue for people with office jobs. Make sure at least every hour, you are taking breaks. If you work from home, try purchasing a desk with an adjustable height (like the ones you see in movies about hipster startups) so that at times, you'll be able to work at it while standing.

Once you understand the concept behind physical movement's positive effect on our lymphatic system, it's easy to understand and to remember to keep your body in motion—even when lymph pain or inflammation tempts you to remain still. Linda T. Miller, PT, DPT, CLT, and clinical director of the Breast Cancer Physical Therapy Center, Ltd., in Malvern, PA explained that every time we move a muscle in our bodies, the action is achieved through a pump and release motion: "The lymphatic vessels lie between the muscle and the skin. With activity, the muscle contracts and relaxes against the skin. So by pump and release, the activity massages the lymph vessels and moves extra lymphatic fluid out of there."

6. Take brisk walks

When you walk, your legs act as a natural pump that helps move lymph fluid through your body. Deep breathing as you walk creates pressure and expansion that also helps circulate your lymphatic system. Swing your arms and power walk for best results.

7. Drink tons of water

Dehydration is a common cause of lymph congestion, making it thick and less mobile. Drink a glass of clean, filtered water eight to ten times per day. It's worth remembering that the lymphatic system itself is composed of about 96 percent water. Obviously, it can't function if the body doesn't have an adequate supply of water in it.

8. Jump on a rebounder

A rebounder is a mini trampoline and one of the most efficient ways to stimulate lymph flow. How does it work? The gravitational pull caused by gentle bouncing causes the one-way lymphatic valves to open and close, moving the lymph. If you don't have a rebounder, use an exercise or yoga ball. It does the same trick!

9. Stretch or practice yoga daily

In Yin Yoga and Restorative Yoga, stretches are held for up to five minutes. Combined with conscious breathing, these held poses can help direct lymph through the deep channels of the chest.

10. Massage your breasts

Breast massage is great for women who have had a mastectomy or have had lymph nodes removed and are experiencing lymphedema.

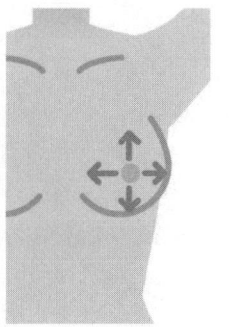
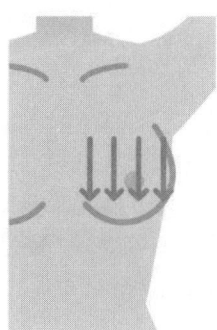
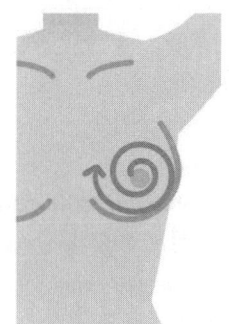

Here are some quick pointers on how to get the most out of a breast massage.

First, sit in front of a mirror and cup your breast with the same hand. Make a "V" shape with the thumb and middle finger of the opposite hand and apply gentle but firm strokes from the outer breast to the nipple. Next, lift your arm up toward the ceiling and again take the opposite hand to massage the upper, outer chest where it connects to the arm and also the inner armpit with gentle strokes. Last, massage under the neck and on the sides of the throat up to the base of the ear.

You may want to use some essential oils that help lymphatic flow. Cleavers and Black Radish are two such oils that Bowden-Luccardi recommends. However, make sure you mix these with a carrier oil such as coconut, jojoba, sweet almond, or rosehip oil to dilute them first.

11. Drink herbal teas that support lymph flow

Red clover, astragalus, mullein, goldenseal, fenugreek, ginger, dandelion, wild indigo root, sarsaparilla, goldenseal, and olive leaf tea are all wonderful for boosting lymph flow.

12. Use paraben- and aluminum-free deodorant

When you shave under your arms, an area that houses many lymph nodes, the pores of your skin are open. Applying deodorant to these open pores lets toxins directly into your body. Make sure you choose a deodorant that is paraben- and aluminum-free.

The Real Cost of Dental Work

One of the first questions I ask a new patient who comes to see me is if they have had any recent dental work done. Very few people realize how common oral inflammation is and its dangers to the rest of your overall health. In fact, your mouth is the perfect place for systemic inflammation to start and spread as it makes its way down the anterior neck and into the lymphatic system.

Doctors are slowly discovering the hidden dangers of root canals and mercury amalgams in cavity fillings. In the 1920s, Dr. Weston A. Price presented research suggesting that bacteria trapped in dentinal tubules during root canal treatment could leak into the body's lymphatic system and cause degenerative systemic diseases such as arthritis and diseases of the kidney, heart, nervous, gastrointestinal, endocrine, and other systems.

Seventy years later, Dr. Boyd Haley studied about nine hundred teeth with root canals for their levels of toxicity. He found that 25 percent of the teeth carried toxins that were fairly benign, roughly 50 percent contained bacteria that would challenge any healthy immune system, and the last 25 percent of the teeth contained highly toxic bacteria.

Dentists who are speaking out against the safety of root canals have three main concerns:

1. There is no way to completely remove all of the dead tissue from the tooth before they place it back into a patient's mouth.
2. There is no way to fully sterilize the tooth and all of its tubules.
3. The materials used to fill the hollowed-out tooth are toxic and leak down into the body's entire system.

The bottom line is: if you have an autoimmune disease, Lyme disease, a heart condition, arthritis, or cancer, root canals could be very risky. If your body is already dealing with these immune challenges and inflammatory responses, then the bacteria from a root canal might push it over a very dangerous edge.

The good news is that new dental technologies are coming onto the scene that are creating safer conditions for patients. Biological dentistry emphasizes the use of nontoxic restorative materials for dental work. This practice treats the teeth, jaw, and related structures, paying specific attention to how the treatment will affect the entire body. I encourage all of my patients, if they can, to switch over to this type of practice.

In the next chapter, I will discuss other healthy practices you should adopt in your daily life to decrease inflammation and toxicity, specifically related to food and nutrition.

What You Need to Know

- Premenstrual Syndrome (PMS) is another fake diagnosis. The real problem is estrogen dominance, a condition that is not normal, even though it is common.
- Chronic stress, poor gut and liver health, and environmental toxins can cause this hormonal imbalance.
- Signs of estrogen dominance include bloating, decreased sex drive, tender and lumpy breasts, mood swings, and weight gain.
- Drinking less alcohol, taking care of the bacterial balance in your gut, boosting your fiber intake, and eating organic and hormone-free are all ways to correct estrogen dominance.
- Leading causes of breast cancer include the average American diet, inflammatory conditions, radiation exposure, and especially environmental endocrine disruptors, such as fossil fuel by-products, toxic chemicals, and environmental pollutants.
- Based on US mortality statistics, screening mammography prevents only one death per 1,000 women screened.
- If you are a woman with dense, fibrous breasts, regular mammography might do very little to protect you.
- A woman who has already developed breast cancer risks rupturing the encapsulation of a cancerous tumor when she gets a mammogram.
- Unlike mammography, thermography works on dense breast tissue and doesn't expose women to harmful radiation.
- Angiogenesis, the increased blood supply to cancer cells, produces measurable amounts of heat that can be detected by thermometric devices.
- Women with dense breast tissue and women under forty are overlooked when it comes to screening for breast cancer.
- Thermograms can identify cancer sooner, with more accurate results.
- There are many ways to take care of your breasts and lymphatic system, including wearing less-constricting bras without

underwire, dry brushing, taking alternating hot and cold water showers, taking brisk walks, moving your body regularly, drinking a lot of water, gently jumping on a rebounder, massaging your breasts, drinking certain herbal teas, and using paraben- and aluminum-free deodorant.
- Root canal and other dental work can spread toxins and cause inflammation throughout the whole body.

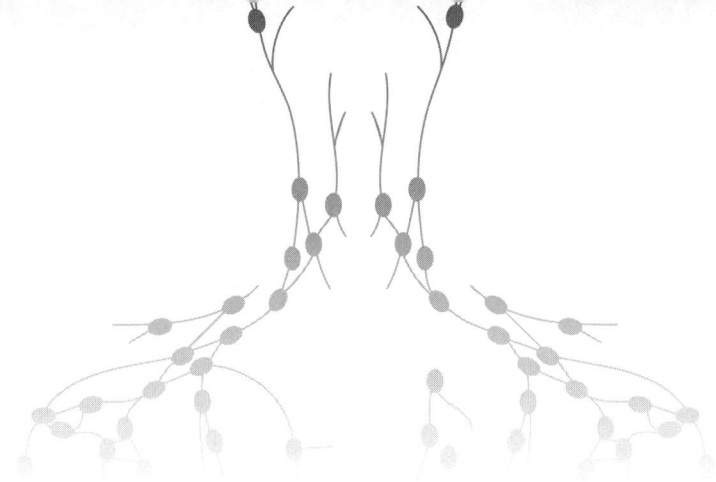

CHAPTER 10

The Lymph Link and Your Diet

Up till now, I've been talking at length about physical toxicity and the many elements in our environment that contribute to it. In this last chapter, I'd like to narrow the focus to the foods you should and shouldn't be eating in order to optimize your lymph health. Scientific research is constantly uncovering new ways food can act as a powerful medicine, whether we are talking about its anti-inflammatory effects, its ability to decrease bad cholesterol, its natural detoxification properties, or the balancing of hormones and other chemicals.

I could—and might—write a whole book on the myriad ways that eating the right foods has helped my patients get back on track with their health. As I have learned over the years, a person's relationship with food is a very personal, lifelong journey that involves listening to your body's responses to what goes into it and then making choices about your diet. These are choices that no one else can make for you as you're walking through the grocery store aisles, ordering dinner at a restaurant, or standing in front of the refrigerator late at night. In this chapter, I will break down for you what I see as the most important nutritional information you can have when it comes to making good decisions about your nutrition that will support your lymphatic system and rid your body of unwanted toxins.

LYMPH-LINK

Your Mom Was Right: Eat Your Veggies (and Fruits)

Let's start with the basics. If you're like many of my patients—and most Americans—you aren't eating enough fruits and vegetables, which contain important antioxidants that break down dangerous free radicals in your cells. In fact, according to a recent study by the CDC, only about one in ten adults meet the federal recommendation of one and a half cups of fruits and two to two and a half cups of vegetables each day. This deficit puts people at higher risk for heart disease, diabetes, and certain cancers. It doesn't matter what you call the disease, or what its symptoms are. As I've said throughout this book, the problem starts with inflammation, and most fruits and vegetables have natural anti-inflammatory and detoxifying effects that you can't afford to do without. Some of the best vegetables you can consume for detoxification are found in the cruciferous family and include broccoli, cabbage, cauliflower, kale, and Brussels sprouts. When digested, these veggies produce a compound called indole-3-carbinol (I3C) that is known to stimulate detoxifying enzymes in the gut and liver.

A large percentage of the immune system, which is the major system associated with inflammation, exists in the gastrointestinal tract. A Mediterranean diet, rich in fruits, vegetables, whole grains, nuts, legumes, and olive oil, is known to reduce inflammation by supporting your GI tract. This diet works by increasing the intake of nutrients such as potassium, calcium, and magnesium—all great inflammation fighters—and at the same time reducing the intake of sodium. The fiber, healthy fats, vitamins, minerals, and polyphenols from plant-based foods in this diet all keep the bacteria in a person's gut healthy and balanced.

In contrast, eating large amounts of red meat, sugar, and processed foods will feed the unhealthy bacteria in your GI tract and lead to an increase in inflammation. Another word about sugar: I advise my patients to keep their sugar intake to twenty to thirty grams or lower a day and even lower if they have a history of diabetes. The liver is one of the three organs responsible for blood sugar balance, so eating too much sugar makes it very difficult for the liver to do its job of detoxification.

DR. LORETTA T. FRIEDMAN

The Dirty Dozen versus the Clean Fifteen

As you saw in Chapter 5, our environment is saturated with toxic chemicals that seem almost impossible to escape. Factory farms that mass produce livestock, seafood, dairy, and produce are some of the worst culprits for chemical pollution, which is why it is so important whenever possible to consume organically-grown foods. I understand that this is not always easy or affordable. The Environmental Working Group has tirelessly been providing yearly updates to a list of most toxic produce, which they call The Dirty Dozen to help consumers prioritize what to buy organic. They also have a list of the safest fruits and vegetables to consume if organically-grown ones are not available.

Prioritize buying these fruits and vegetables organic if and when you can:

1. Strawberries
2. Spinach
3. Kale, collard, and mustard greens
4. Nectarines
5. Apples
6. Grapes
7. Cherries
8. Peaches
9. Pears
10. Bell and hot peppers
11. Tomatoes
12. Celery

The following choices contain the least amount of pesticides, so buying them organic is less important:

1. Avocados
2. Sweet corn
3. Pineapple
4. Onions

5. Papaya
6. Sweet peas (frozen)
7. Eggplant
8. Asparagus
9. Broccoli
10. Cabbage
11. Kiwi
12. Cauliflower
13. Mushrooms
14. Honeydew melon
15. Cantaloupes

Super Foods that Help You Go with the Lymph Flow

Since movement acts as a pump for the lymphatic system, you can't go wrong with foods that promote motion within the body. As you already know by now, your lymphatic system plays one of the most important roles in cleansing your body. It keeps your blood healthy, your digestive system humming, and helps the body detox from harmful substances the way that nature intended it to do. However, as you've also seen, toxins can slow down the lymph. Besides avoiding dairy, processed foods, sugary foods, and foods high in oils, all of which slow the lymph down, you should consume large quantities of foods that energize. The following foods will help energize you so that you are more active and help the natural flow of the lymph.

1. Citrus

Citrus is one of the most detoxifying foods you can eat, specifically lemons, limes, navel oranges, tangerines, blood oranges, and grapefruit. These foods all possess powerful enzymes, along with Vitamin C, that support the body and keep digestion flowing. Enjoy one to three servings of these daily, whether it's the whole fruit, the juiced version, or squeezed onto salads or entrees.

2. Berries

Berries, especially cranberries, are rich in detoxifying properties that cleanse the system and also add hydration for healthy lymph flow. Since lymph helps the body filter toxins, it's important to eat foods with anti-bacterial benefits such as cranberries.

3. Greens

Greens remove harmful chemicals and toxins from the body that we encounter daily. They also provide nutritional support with Vitamins A, C, and K, along with iron, magnesium, B vitamins, and protein.

4. Sunflower and Pumpkin Seeds

These seeds provide magnesium to support the nervous system and healthy fats to lubricate the body and promote lymph flow. They also contain fiber, which helps scrape the digestive tract to prevent toxic-buildup that can block lymph flow. Toss a handful of these seeds into your next bowl of porridge or salad or use them to bake with.

5. Chia, Hemp, and Flax

Like pumpkin and sunflower seeds, these seeds are full of Omega-3—rich fatty acids, which are excellent for lymph flow. Hemp seeds are also a good source of chlorophyll, nature's best cleanser, so be sure to include some in your daily diet. Their fiber also keeps the body's processes moving and flowing naturally and will help your heart stay healthy. Enjoy them in oatmeal or sprinkle them into a smoothie or entree. Lastly, healthy fats like these help your body absorb nutrients from other foods that the lymphatic system needs to work well.

6. Herbs and Spices

All herbs and spices are very useful for cleansing the body. Turmeric, ginger, cinnamon, cardamom, coriander, and black pepper are some of the best in this regard. These foods also contain antioxidants that benefit the brain and can help the digestive system work well, again, benefiting lymph flow.

7. Seaweed and Algae

Spirulina, kelp, nori, wakame, chlorella, and dulse are types of seaweed and algae that are not just cleansing for the body, but also incredibly nutrient-dense, providing high amounts of iron, Omega-3s, Vitamin A, protein, magnesium, B vitamins, iodine, and chlorophyll. Try to add a teaspoon of these into your daily diet, if you can, by adding them to your smoothies, salads, and wraps.

8. Avocados

Over the past few years, avocados have flooded into supermarkets and restaurants as more people have caught wind of their health-giving benefits. Avocados are rich in glutathione, an antioxidant that helps prevent cell damage caused by free radicals and heavy metals. A quarter of an avocado per day is okay. Glutathione also supports phase one and phase two of liver detoxification, which I talked about in Chapter 1.

Daily Lymph Link Food Hacks

In addition to the above superfoods, there are a few more nutrition-based routines you can implement each day to help promote lymph health and detoxification.

Garlic

Not only does it help to ward off vampires, but garlic also lowers cholesterol and blood pressure and helps fight diabetes. And that's not all: garlic also promotes lymph flow. Since most of garlic's benefits come when it's in its raw state, the best way to consume it is to chop a clove into quarters and swallow the pieces whole with a glass of water.

Lemon Water

Lemons contain twenty-two anti-cancer compounds, including limonene, an oil that has been shown to halt the growth of cancer tumors in animals. In addition, they also contain flavonol glycosides, compounds known to stop the division of cancer cells. Lemons are considered one of the most alkalizing foods a person can eat, a fact that may seem counterintuitive, since the fruits are acidic on their own. However, in the body, lemons become alkaline once metabolized and in turn help to alkalize the blood, drawing uric acid from the joints and reducing the pain from inflammatory conditions such as arthritis.

Because lemon juice is similar in atomic composition to digestive juices and saliva, it does an excellent job in breaking down material and encouraging the liver to produce bile during phase one of liver detoxification. In addition, the high amounts of Vitamin C found in lemons neutralize free radicals both inside and outside of cells. It also helps prevent cholesterol buildup along the artery walls and consequently helps to stop the progression of atherosclerosis and heart disease.

Turmeric

Turmeric is a bright orange-yellow spice included in many South Asian foods. Curcumin is the active component that gives this spice its bright color and its anti-inflammatory properties. For centuries, many countries around the world have used turmeric medicinally, and alternative medicine practitioners in the states have also caught onto it. In fact, many non-traditional doctors prescribe the spice instead of ibuprofen for ailments such as joint pain, rheumatoid arthritis, depression, diabetes, and irritable bowel syndrome. Try to use turmeric in your cooking or make a tea with it, whether using powder or the raw root. This could be a healthy alternative to that second cup of coffee you're used to making mid-morning.

Foods to Avoid

Dairy

Dairy is known to increase inflammation in the body, so it's a good idea to limit your intake of it as much as possible. Luckily, grocery stores and restaurants offer many alternatives to cow's milk-based products, such as oat milk, almond milk, and soy milk.

Farm-Raised Seafood

The rise of fish farms throughout the world is, in my opinion, a huge atrocity to nature and our health. Many Asian countries have stripped their waters of their beautiful mangrove forests to build giant fisheries that mass produce at alarming rates. Normally, it takes about a year for a medium-sized shrimp to reach full growth. However, these manmade hatcheries are pumping enough chemicals and steroids into the water to get these shrimp to grow to full size in just three months. These farms are severely overcrowded so that the fish hardly have room to move and are constantly swimming in and consuming their own waste.

The toxicity levels in farmed seafood alone should alarm you. But what's just as devastating is the permanent damage these farms are doing to our oceans and waterways, turning the areas they occupy into wastelands where nothing can grow after only eighteen to twenty-four months.

Caffeine

Caffeine is perhaps the most widely-used and accepted drug out there. However, it creates a rollercoaster of fatigue and stress within your adrenal system. If you drink more than one cup (containing seventy to ninety-five milligrams of caffeine) a day, I recommend that you replace some of it with green tea, which contains only twenty milligrams of caffeine.

What You Need to Know

- Only about one in ten adults meet the federal recommendation of one and a half cups of fruits and two to two and a half cups of vegetables each day.
- A Mediterranean diet, rich in fruits, vegetables, whole grains, nuts, legumes, and olive oil, is known to reduce inflammation by supporting your GI tract.
- I advise my patients to keep their sugar intake to twenty to thirty grams or lower a day.
- The best detoxifying/anti-inflammatory superfoods include citrus, berries, avocados, sunflower seeds, hemp seeds, flax seeds, and greens.
- Lemon water helps fight cancer and alkalinizes the blood.
- Turmeric has amazing anti-inflammatory properties that have been recognized for years.
- Dairy, farm-raised seafood, and caffeine are among the foods you should avoid when trying to detoxify and balance your immune system.

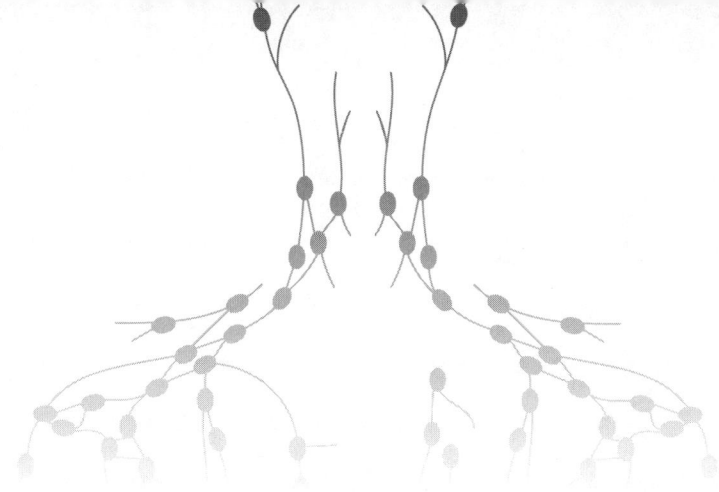

AFTERWORD

These nutritional guidelines are really just the tip of the iceberg. In fact, so is all of the information I have provided here in this book. The bottom line is all of us must take charge of our own journeys back to health. This means listening to your body more than your doctor sometimes. It also means sacrificing your time, energy, comfort level, and finances. It means not turning a blind eye to the information you have just learned throughout these chapters, but instead making choices that are based on your new knowledge.

I have devoted most of my life and career to helping people who are suffering on a daily basis, many of whom have beat the odds and returned from life-threatening disease. I know that the information I have shared with you here in this book works because I have seen it work. Please feel free to visit my website at www.synergyhealthassociates.com, where you can also book a consultation call and learn more.

In the meantime, my greatest wish is that the information I have shared about patients just like you has helped solve some of the mysteries behind your suffering and encouraged you to continue on the path of healing. As I have learned personally, relief is always just around the corner.

Wishing you many pain-free days ahead,
Dr. Loretta T. Friedman

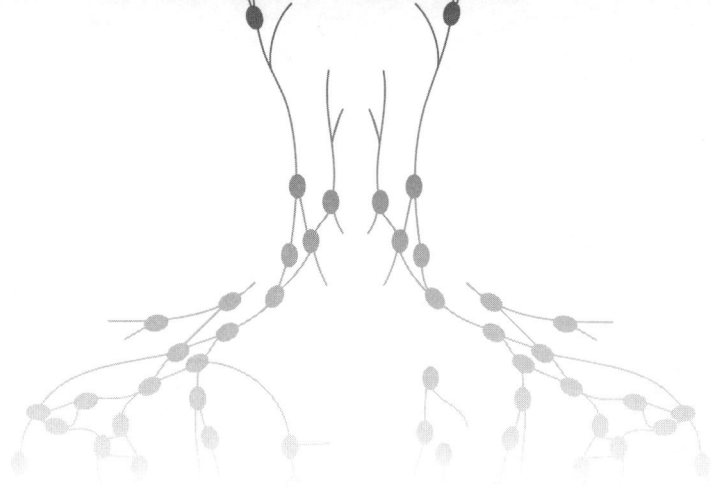

GLOSSARY

Adrenal glands—parts of the body that produce a handful of hormones that help maintain salt balance in our blood and tissues, maintain blood pressure, and produce some sex hormones.

Alkaloid—a type of chemical found in plants that often acts as a drug or poison or is used in medicines.

Allergy—an abnormally-high sensitivity to certain substances such as pollens, foods, or microorganisms. Common indications of allergy may include sneezing, itching, and skin rashes.

Anatomy—the science of the shape and structure of organisms and their parts.

Anesthetic—a substance that makes you unable to feel pain.

Antibody—a protein produced in the blood that fights diseases by attacking and killing harmful bacteria, viruses, etc.

Antioxidant—a substance that may prevent or delay some types of cell damage that occurs through oxidization.

Ashwagandha—a preparation usually of the leaves or roots of an evergreen shrub (Withania somnifera) native to Africa, Asia, and southern Europe that is used in herbal medicine; especially as a tonic, anti-inflammatory, and adaptogen.

Asthma—a condition in which your airways narrow, swell, and may produce extra mucus. This can make breathing difficult and trigger coughing, a whistling sound (wheezing) when you breathe out, and shortness of breath.

Autoimmune disease—any of a large group of diseases characterized by abnormal functioning of the immune system that causes your immune system to produce antibodies against your own tissues.

Artery—a vessel through which the blood passes away from the heart to various parts of the body.

Arthritis—inflammation of any joint.

Bacteria—microscopic living organisms, usually one-celled, that can be found everywhere. They can be dangerous, such as when they cause infection, or beneficial, as in the process of fermentation (such as in wine) and that of decomposition.

Bile—a bitter, alkaline, yellow or greenish liquid, secreted by the liver, that aids in absorption and digestion, especially of fats.

Biotoxin—any toxin produced by a living organism.

Bloodstream—the blood flowing through the circulatory system and the path that it takes.

Blood vessels—components of the circulatory system, such as arteries, capillaries, or veins, that carry blood.

BPA—bisphenol A, an industrial chemical that has been used to make certain plastics and resins since the 1950s.

Breast cancer—cancer of the breast; one of the most common types of cancer in women in the US.

Cancer—the name for a group of more than one hundred diseases in which cells begin to grow out of control. Cancer can develop anywhere in the body. It starts when cells grow out of control and crowd out normal cells.

Cannabis—a drug, illegal in many countries, that is made from the dried leaves and flowers of the hemp plant. Cannabis produces a pleasant feeling of being relaxed if smoked or eaten. Also called marijuana.

Carcinogen—any substance that produces cancer.

Cell—the basic unit of life. All living organisms are either single cells or are multicellular organisms composed of many cells working together. Cells are the smallest known unit that can accomplish all of these functions.

Cell test—a test that examines the functions of the cells to determine if they have been affected by the presence of toxins.

Chemical—anything which is made of matter and has mass; any solid, liquid, or gas.

Chemotherapy—a drug treatment that uses powerful chemicals to kill fast-growing cells in your body. Chemotherapy is most often used to treat cancer, since cancer cells grow and multiply much more quickly than most cells in the body.

Chiropractic—a health-care profession that focuses on the spine and other joints of the body and their connection to the nervous system.

Chronic—lasting for a long period of time or marked by frequent recurrence as certain diseases.

Comorbidity—the presence of two or more conditions occurring in a person either at the same time or successively (one condition that occurs right after the other). Conditions described as comorbidities are often long-term (chronic) conditions.

Complex decongestive therapy—the current standard therapeutic intervention against lymphedema. It involves manual lymph drainage by a physician and the frequent wearing of compression garments.

Compression garment—elastic clothing with an engineered compression gradient that can be worn on limbs, upper, lower, or full body to use for therapy and sports.

Cortisol—the body's main stress hormone, created by the adrenal system when we undergo stress.

CYP450 (Cytochrome P450)—an enzyme normally used to fight free radicals, which contribute to inflammation and disease.

Debulking—removal of a major portion of the material that composes a lesion, such as the surgical removal of most of a tumor so that there is less tumor load for subsequent treatment by chemotherapy or radiation.

Dental reconstructive surgery—rebuilding and/or replacing the teeth in a patient's mouth.

Detoxification—the process by which toxins are changed into less toxic or more readily-excretable substances.

Diabetes—a disease that occurs when your blood glucose, also called blood sugar, is too high.

Diagnosis—the act of identifying a disease, illness, or problem by examining someone.

Diet—food and drink regularly consumed by a person.

Dietary sensitivity—food sensitivity.

Directional Non-Force Technique Chiropractic—a low-force method of chiropractic that utilizes a diagnostic system involving a unique type of leg check.

Disease—any condition that prevents the body or mind from working normally.

Disinfection—treatment with disinfectant materials in order to destroy harmful microorganisms.

Elective surgery—surgery that is not essential, especially surgery to correct a condition that is not life-threatening; surgery that is not required for survival.

Electrostatic field—electric field associated with static electric charges.

Endocrine system—a network that uses hormones to control and coordinate your body's metabolism, energy level, reproduction, growth, development, and response to injury, stress, and mood.

Enterohepatic circulation—the circulation of biliary acids, bilirubin, drugs, or other substances from the liver to the bile, followed by entry into the small intestine, absorption by the enterocyte, and transport back to the liver; a problem with the waste removal system.

Environmental toxin—a small amount of poison found in air, water, food, etc.

Epidemiology—the branch of medical science that investigates all the factors that determine the presence or absence of diseases and disorders.

Essiac tea—a drink introduced to the world in 1922 by a nurse named Rene Caisse; used to fight cancer.

Failed back surgery syndrome—a term that is often used to describe the condition of patients who have not had a successful result with back surgery or spine surgery and have experienced continued pain after surgery.

Fluoride—a chemical substance sometimes added to water or toothpaste (= substance for cleaning teeth) in order to help keep teeth healthy.

Food sensitivity—a diffuse and poorly-understood reaction to food that may be associated with increased levels of certain IgG class antibodies that are reactive to that food; an adverse reaction to a food in the diet that is separate from an allergy and does not involve the immune system.

Formaldehyde—a chemical used in many industries, including glues, resins, disinfectants, dyes, textiles, car parts, labs, and for preserving dead bodies and body parts.

Free radical—a molecule that has an extra electron and therefore reacts very easily with other molecules.

Functional medicine—a systems biology-based approach to health that focuses on identifying and addressing the root cause of disease. Each symptom or differential diagnosis may be one of many contributing to an individual's illness.

Fungus—a simple organism, or living thing, that is neither a plant nor an animal. Fungal infections come in different forms, like ringworm, athlete's foot, toenail fungus, yeast infections, and jock itch.

Gastroenterology—the branch of medicine that deals with the diagnosis and treatment of diseases and disorders of the digestive system.

Generalized—involving the whole of an organ as opposed to a focal or regional process.

Genetics—the science of heredity: what you get from your parents and ancestors in your DNA.

Glyphosate—a chemical that is used to kill weeds.

Heart disease—a structural or functional abnormality of the heart, or of the blood vessels supplying the heart, that impairs its normal functioning.

Heavy metals—metals with high densities, atomic weights, or atomic numbers. They can include lead, mercury, arsenic, zinc, copper, aluminum, thallium, and cadmium. They can be dangerous toxins when found in the human system.

Hepatic malfunction—when the liver does not work properly.

Hodgkin Lymphoma—a form of lymph cancer in which cells in the lymphatic system grow abnormally and may spread beyond it.

Holistic medicine—an approach to medical care that emphasizes the study of all aspects of a person's health, including physical, psychological, social, economic, and cultural factors.

Hormone—a substance that is produced by the body and stimulates certain activity in the cells.

Hormone imbalance—when there is too much or too little of some hormone in the body.

Immune system—the complex system throughout the human body that provides a general defense against harmful germs, substances, and infections.

Increased intestinal permeability—leaky gut syndrome.

Infection—occurs when another organism enters your body and causes disease. The organisms that cause infections are very diverse and can include things like viruses, bacteria, fungi, and parasites.

Inflammation—a localized protective response elicited by injury or destruction of tissues, which serves to destroy, dilute, or wall off both the injurious agent and the injured tissue.

Insomnia—the inability to sleep.

Insulin—a hormone in the body that controls the amount of sugar in the blood.

Kratom—an evergreen tree native to Thailand. The leaves are boiled to release an intoxicating sedative and stimulant.

Leaky gut syndrome—a gastrointestinal condition that some have proposed contributes to a range of whole-body health problems, such as irritable bowel syndrome, skin rashes, chronic fatigue syndrome, and mood disorders.

Lesion—any pathological or traumatic discontinuity of tissue or loss of function of a part.

Liver—the main organ of the human body responsible for processing toxins.

Low-level laser therapy—a form of medicine that applies low-level (low-power) lasers or light-emitting diodes (LEDs) to the surface of the body.

Lumbar support—physical bolstering of one of the most important vertebrae in the back.

Lyme disease—an acute inflammatory disease that is transmitted by ticks.

Lymph—a clear, watery fluid derived from body tissues that contains white blood cells and circulates throughout the lymphatic system, returning to the venous bloodstream through the thoracic duct. Lymph acts to remove bacteria and certain proteins from the tissues, transport fat from the small intestine, and supply mature lymphocytes to the blood.

Lymphatic blockage—any obstruction in the lymphatic system.

Lymphatic drainage massage—when a practitioner uses a range of specialized and gentle rhythmic pumping techniques to move the lymph fluid in the direction of the lymph pathways.

Lymphedema—a build-up of lymph fluid in the fatty tissues just under your skin. This build-up might cause swelling and discomfort. It often happens in the arms or legs, but can also happen in the face, neck, trunk, abdomen (belly), or genitals.

Lymph nodes—an important part of the immune system, acting as "nodes" between the lymphatic vessels that span the body. Immune cells that cluster in these nodes stand ready to attack any bacteria, viruses, or other foreign substances that enter the body.

Lymph massage—a form of gentle massage that encourages the movement of lymph fluids around the body.

Lymphoma—cancer that begins in cells of the lymph system.

Mainstream medicine—referring to conventional (i.e., non-alternative or non-complementary) medicine or medical practice; assumes that all physiologic and pathological phenomena can be explained in concrete terms, and "best practice" is the end result of a stream of objective analyses.

Mammogram—an x-ray picture of the breasts (mammaries), used to screen for breast cancer.

Mastectomy—surgical removal of all or part of the breast and sometimes associated lymph nodes and muscles.

Metabolism—all the chemical reactions that occur in a living organism to sustain life.

Microplastic—an extremely small piece of plastic that is harmful to the environment, used in cosmetics, or formed when plastic waste material breaks down.

Microsurgery—dissection of minute structures under the microscope with the use of extremely small instruments.

Misdiagnosis—a decision that a person has a particular illness or condition, when in fact they have a different one, or none at all.

Moxa—a Japanese heat treatment that ignites small cones made of mugwort.

Nanoparticle—a microscopic particle whose size is measured in nanometers.

Neogenesis—the regeneration of tissue.

Non-Hodgkin Lymphoma—lymph cancer in which white blood cells called lymphocytes grow abnormally and can form growths (tumors) throughout the body.

Obesity—a condition involving an excessive amount of body fat.

Omega-3—a substance in the oil from some fish such as tuna and salmon and in some seeds thought to be good for your health.

Oscillation—a backward and forward motion, like that of a pendulum; also vibration, fluctuation, or variation.

Oncology—a branch of medicine concerned with the prevention, diagnosis, treatment, and study of cancer.

Ozone—O3, a modification of oxygen, having increased chemical activity; a colorless gas having a peculiar odor like that of air, which contains a trace of chlorine.

Ozone therapy—the process of administering ozone gas into your body to treat a disease or wound.

Patellofemoral Pain Syndrome—the name given to most knee and leg pain.

PCBs (Polychlorinated Biphenyls)—a family of highly toxic chemical compounds; known to cause skin diseases and suspected of causing birth defects and cancer.

Pesticide—substance or agent used to kill pests such as unwanted or harmful insects, rodents, or weeds.

Phenol—a caustic, poisonous, white crystalline compound used in resins, plastics, and pharmaceuticals and in dilute form as a disinfectant and antiseptic.

Physician—a person trained and licensed to practice medicine; a medical doctor.

Physiology—the biological study of the functions of living organisms and their parts.

Placebo—a substance given to someone who is told that it is a particular medicine either to make that person feel as if they are getting better or to compare the effect of the particular medicine when given to others.

Plastic surgery—surgery to remodel, repair, or restore the appearance and sometimes the function of body parts. It includes reconstructive surgery, such as skin grafts, and repair of congenital defects as well as cosmetic surgery.

Pneumatic compression pump—device that administers intermittent compression via an inflatable garment, such as an arm sleeve, pant leg, or vest. Some patients experience a reduction of fluid and swelling after a session on a pneumatic compression device.

Radiation—energy that comes from a source and travels through space at the speed of light. This energy has an electric field and a magnetic field associated with it and has wave-like properties. You could also call radiation "electromagnetic waves."

Sciatic pain—pain that radiates along the sciatic nerve and is typically felt in the buttocks, down the back of the leg, and possibly to the foot.

Shiatsu massage—a technique that involves manual pressure applied to specific points on the body to relieve tension and pain.

Single nucleotide polymorphisms—the most common type of genetic variation, which occurs throughout all of our DNA.

Spleen—lymphoid organ, lying in the human body to the left of the stomach below the diaphragm, serving to store blood, disintegrate old blood cells, filter foreign substances from the blood, and produce lymphocytes.

Steatohepatitis—a form of liver inflammation characterized by an accumulation of fat.

Sterilization—the process of treating something to kill or inactivate microorganisms.

Subluxation—a misalignment of tissue in the body.

Supplement—something added to a food or a diet to increase its nutritional value.

Surgery—the treatment of injuries or diseases in people or animals by cutting open the body and removing or repairing the damaged part.

Swelling—the enlargement of organs, skin, or other body parts. It is caused by a buildup of fluid in the tissues.

Symptom—An indication of a disorder or disease.

Systemic—relating to or affecting the entire body or an entire organism.

Temporomandibular joint dysfunction (TMJD)—When a person experiences pain in the joints that act as a hinge for the jaw, connecting it to the rest of the skull, and/or the muscles that control them.

Thermography—a technique wherein an infrared camera photographically portrays the body's surface temperature based on self-emanating infrared radiations; used as a diagnostic aid in the detection of breast tumors and the assessment of rheumatic joints; also used in the study of pain.

Thoracic Outlet Syndrome—pain going down your arm or in your forearm or fingers, which may be due to a misaligned disc in your neck or T1 rib.

Thyroid—the gland which tells your body how fast to tick over by generating the known hormone called thyroxine.

Toxicity—the quality of being poisonous or the presence of a toxin.

Toxin—a poison; any substance which does harm when introduced to the body.

Trichloroethylene—a solvent and as a chemical used to make other chemicals; dangerous and found in many parts of the human environment.

Ulcer—a sore on the lining of your stomach, small intestine, or esophagus.

Vascular system—the vessels and tissue that carry or circulate fluids such as blood or lymph through the body.

Vein—one of the systems of branching vessels or tubes conveying blood from various parts of the body to the heart.

Virus—a microorganism that is smaller than a bacterium that cannot grow or reproduce apart from a living cell. A virus invades living cells and uses their chemical machinery to keep itself alive and to replicate itself.

Volatile organic compounds—chemicals that have a high vapor pressure and low water solubility; properties that help them stick around in our air and water supplies for a long time.

Water pill—any substance that tends to increase the flow of urine, which causes the body to get rid of excess water.

White blood cells—blood cells that engulf and digest bacteria and fungi; an important part of the body's defense system.

Western medicine—the typical methods of healing or treating disease that are taught in western medical schools, using hypothetical deduction, in which doctors see patients and treat their symptoms with the use of prescription medications, surgical operation, various forms of therapy, and radiation.

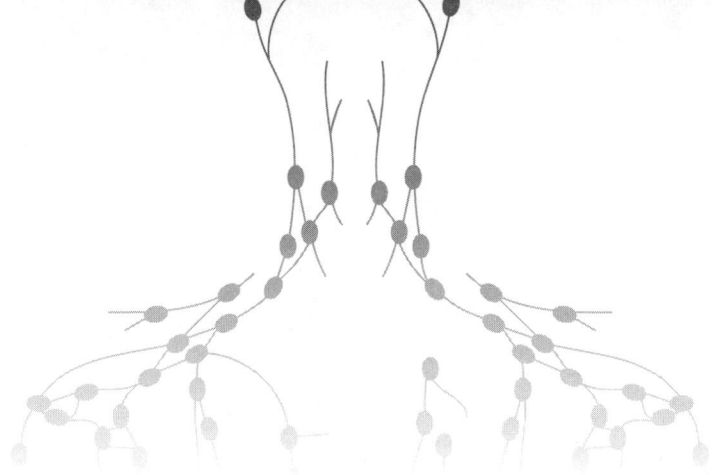

METABOLIC DETOXIFICATION RECIPES THAT WILL MAKE YOUR MOUTH WATER

A few notes to take into consideration:

- Ingredients should be organic when possible.
- I recommend that all meats and chicken be grass-fed or grass-finished.
- Fish should be wild-caught, not farmed.
- Recipes can be doubled to increase the amount of servings.
- Salad dressings are good in the refrigerator for three days.

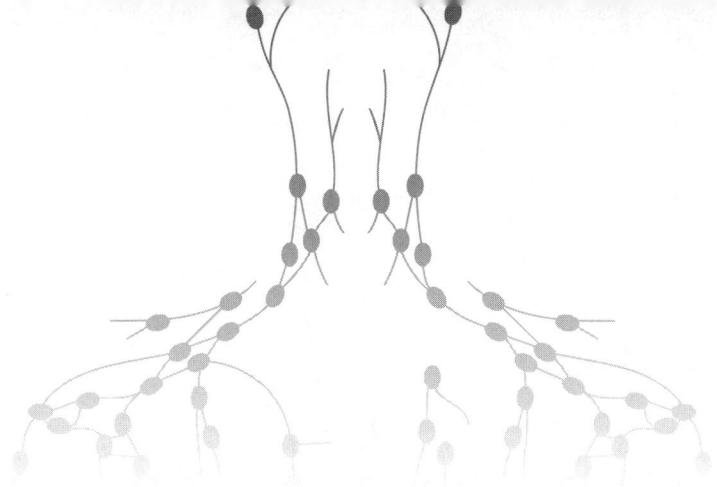

TEAS AND SMOOTHIES

Ginger Turmeric Tea
Lavender Peach Tea
Beet Lime Ginger Juice
Orange Tea
Green Ginger Smoothie
Lemon Green Smoothie

Ginger Turmeric Tea (Concentrated)

Yield: 2 servings
Serving Size: 2 tablespoons
Prep Time: 5 minutes
Cook Time: 30 minutes

Ginger has been known to help with digestion, and turmeric is a great tubular for helping with detoxing the body because of the compound called curcumin. Both ingredients contribute to helping fight inflammation. For additional detoxing properties, add mint and/or lemon. This easy tea can be made ahead of time and refrigerated for up to two weeks.

Ingredients

4 cups of filtered water
1 cup fresh ginger with skin on, sliced into 1-inch pieces
1 cup fresh turmeric with skin on, sliced into 1/4-inch pieces
Lemon—optional
Mint—optional

Directions

1. Wash the ginger and turmeric thoroughly and slice them into pieces.
2. Place the ginger and turmeric slices into a pot of 4 cups of filtered water.
3. Bring the water to a boil with the ginger and turmeric. Once boiling, lower the heat.
4. Allow to simmer for about 30 minutes, or until half the liquid has evaporated.
5. At this point, turn off the heat and strain the mixture, keeping the liquid. You should end up with about 2 cups of concentrated ginger turmeric tea.

For a hot tea, add 2 tablespoons to hot water. Add a squeeze of lemon (optional).

This mixture can be added to filtered cold water or sparkling water too for an iced tea.

Nutritional Information: Calories 10, Fat 0 g, Sodium 0 mg, Carbohydrates 2 g, Fiber 0 g, Sugar 0 g; Protein 0 g.

Lavender Peach Tea

Yield: 2 servings
Total Time: 10–12 minutes

A fruity tea that is both refreshing and purposeful, the ingredients in this fruity tea will help cleanse the digestive tract and are beneficial for detoxing throughout the lymphatic system.

Ingredients

1 peach, fresh, pitted, diced
1/4 cup basil, leaves, fresh
1 tablespoon lavender, dried
1" piece ginger, fresh, peeled, sliced
4 cups water

Directions

1. Place diced peach, fresh basil, dried lavender, and sliced ginger into a small saucepan.
2. Cover with water and place over medium-high heat to bring to a boil.
3. Once boiling, remove saucepan from heat and cover to steep for 4–6 minutes.
4. Strain tea through a fine mesh strainer into mugs and enjoy.

Nutritional Information:
Calories 25, Fat 0 g, Sodium 20 g, Carbohydrates 6 g, Fiber 1 g, Sugar 5 g; Protein 1 g.

Beet Lime Ginger Juice

Yield: 1 serving
Total Time: 5 minutes

Beet juice is an excellent detoxifier of the liver. Beets contain betalains that are shown to be anti-inflammatory, help to increase bile, and activate liver enzymes. This helps to protect the liver from damage and keep detox pathways in check.

Ingredients

3 medium beets, peeled
1 lime, peeled
1" piece of ginger, peeled
Juicer or blender

Directions for juicer

1. Chop beets into large chunks.
2. Juice all ingredients according to your juicer instructions and enjoy over ice.

Directions for blender

1. Chop beets into large chunks and add to a blender with the lime and ginger.
2. Blend until smooth. Strain mixture through a fine mesh strainer lined with cheesecloth. Enjoy over ice.

Nutrition Information:
Calories 130, Fat 0 g, Sodium 0 g, Carbohydrates 32 g, Fiber 9 g, Sugar 18 g; Protein 5 g.

Orange Tea

Yield: 2 servings
Total Time: 10–12 minutes

Orange tea helps boost overall body immunity by increasing vitamin C intake. It may also relieve respiratory problems associated with a cold, flu, or asthma by expelling congestion. Cardamom and cinnamon are anti-inflammatory, contain many antioxidants, and may help prevent cancer.

Ingredients

1 medium orange
1 cinnamon stick
5–6 cardamom pods
4 cups water
Honey (optional)

Directions

1. With a vegetable peeler, peel the orange, trying to avoid the white pith. You should have several 3- inch peel slices.
2. Add the orange peel to a small saucepan with cinnamon stick and cardamom pods and top with water.
3. Bring mixture to a boil and then remove from heat and cover. Allow to steep for 6 minutes.
4. Strain mixture through a fine mesh strainer and serve in mugs.
5. Sweeten with honey, if desired.

Nutritional Information:
Calories 40, Fat 0 g, Sodium 20 mg, Carbohydrates 10 g, Fiber 3 g, Sugar 7 g; Protein 1 g.

Green Ginger Smoothie

Yield: 1 serving
Total Time: 5 minutes

This smoothie packs a punch with nutrients and antioxidants. Spirulina contains chlorophyll that helps to remove toxins from the body. Research has shown that spirulina has anti-inflammatory properties and pain relief effects as well as brain-protective properties.

Ingredients

1 banana, frozen
1 cup spinach, fresh
1/2 cup zucchini, raw, diced, frozen
1 tablespoon spirulina powder
1" piece ginger, fresh, peeled
1 tablespoon pepita (pumpkin) seeds, raw
1 cup almond milk

Directions

1. In a blender, add the frozen banana, spinach, zucchini, spirulina powder, ginger, pepita seeds, and almond milk.
2. Blend until fully incorporated, then add to a glass and enjoy.

Nutritional Information:
Calories 310, Fat 7 g, Sodium 280 mg, Carbohydrates 52 g, Fiber 7 g, Sugar 24 g; Protein 13 g.

Lemon Green Smoothie

Yield: 1 serving
Prep Time: 5 minutes
Cook Time: 1 minute

This dish is great for detoxing the liver with the aid of the sulfur-containing broccoli sprouts. It also boasts great flavor with detoxing lemon and pineapple, with an added boost of vitamins and antioxidants from the barley grass juice powder.

Ingredients

2 cups of spinach
1 cup of fresh cilantro with stems
1 thumb of ginger
1 apple
1 cup filtered water
1 cup frozen pineapple
1 tsp barley grass juice powder
1/4 cup of broccoli sprouts
2 tbsp of fresh lemon juice

Directions

1. Roughly chop apple.
2. Peel and chop ginger.
3. Place all ingredients in a high-speed blender, blend, and enjoy!

Nutritional Information:
Calories 200, Fat .5 g, Sodium 75 mg, Carbohydrates 49 g, Fiber 8 g, Sugar 33 g; Protein 4 g.

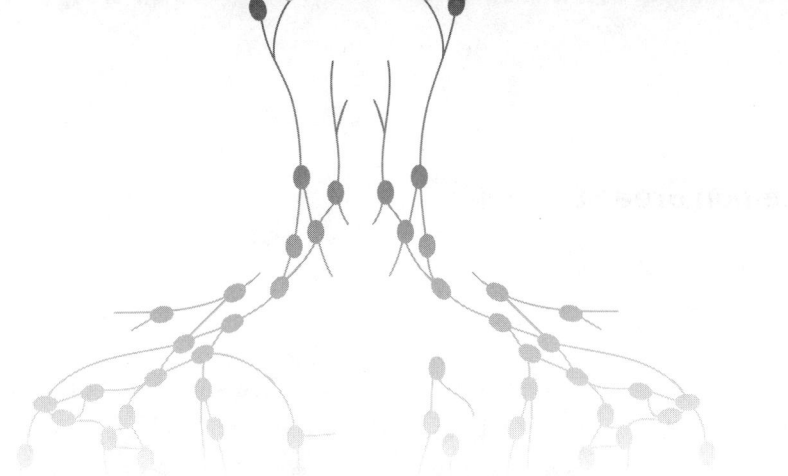

BREAKFAST

Overnight Chia N' Oats
Buckwheat Pancakes
Purple Sweet Potato Toast
Tofu Scramble

Overnight Chia N' Oats

Yields: 1 serving
Time: 5 minutes

This is an easy breakfast, especially if you need something quick in the morning. You can put it together the night before and let it sit overnight. The combination of chia and flax seeds with the berries make this a dynamite detox-er to start your morning.

Ingredients

1/2 cup gluten-free oats
1 tablespoon Chia seeds
1 teaspoon ground flax seed
1/2 cup Almond milk
1 teaspoon Cinnamon
1/4 cup berries of your choice

Directions

1. Place first 5 ingredients in a bowl or canning jar and mix well, cover, and let sit for a minimum of 2 hours or overnight.
2. Top with berries of your choice and add another sprinkle of cinnamon.

Nutritional Information:
Calories 310, Fat 9 g, Sodium 80 mg, Carbohydrates 50 g, Fiber 13 g, Sugar 11 g; Protein 9 g.

Buckwheat Pancakes

Yield: Serves 2
Serving Size: roughly 3–4 pancakes each
Prep Time: 5 minutes
Cook Time: 10 minutes

What is so great about these pancakes? Buckwheat contains protein and fiber. There are probiotics in the buttermilk, which help with digestive issues. Flaxseed supports your heart by lowering cholesterol. All these ingredients help with the natural process of detoxing in the body and are naturally gluten free. However, be sure to use a baking powder that does not contain aluminum.

Ingredients

1 cup buckwheat flour
1 teaspoon baking powder
1 teaspoon baking soda
2 tablespoons ground flax seed
1/4 teaspoon salt
1 cup whole-fat buttermilk
1 egg
½ teaspoon vanilla extract
1 tablespoon raw honey, plus one teaspoon
1 teaspoon oil

Directions

1. In one bowl, combine flour, baking powder, baking soda, ground flaxseed, and salt. Stir to blend well.
2. In a separate bowl, combine the wet ingredients—buttermilk, egg, vanilla extract, and honey. Use a fork to be sure the egg is broken up and the honey is mixed in.

3. Create a well in the dry ingredients and slowly add the wet ingredients, mixing as you are pouring them into the dry ingredients.
4. Heat oil in a pan. Once hot, add 1/4 cup of the mixture into the pan. Cook the pancake for about 3 minutes on one side and then turn over to finish cooking about 2 more minutes. Depending on the size of the pan, you can cook a few pancakes at a time.

Tip: Add black or blueberries to the pancakes to increase the antioxidants.

Nutritional Information:

Calories 490, Fat 12 g, Sodium 1010 mg, Carbohydrates 78 g, Fiber 20 g, Sugar 18 g; Protein 16 g.

Purple Sweet Potato Toast

Yield: Serves 2
Serving Size: 2 slices
Prep Time: 5 minutes
Cook Time: 20 minutes

Whether you eat this for a snack or for breakfast, the sweet potato promotes gut health, is high in fiber and antioxidants, and aids in detoxing. The avocado topping is a bonus in aiding with gut health and detoxing.

Ingredients

1 large purple sweet potato (another color sweet potato can be substituted)
1 avocado, smashed with fork
1/2 teaspoon lemon juice, fresh
Pea shoots or microgreens
Salt and pepper

*Additional topping ideas (optional)

Sliced olives on top of the avocado
Almond butter and banana
Tomato and avocado slices

Directions

1. Preheat oven to 400°F.
2. Wash and dry the sweet potato; slice potato lengthways about 1/4 inch thick.
3. Place potatoes on a sheet pan that is topped with a cooling rack. Place the potatoes on the cooling rack and cook for about 15 minutes in the heated oven. Note: Potato will not be cooked all the way when you take them out.

4. While potato is baking, take avocado and smash it in a bowl. Add lemon juice and mix into the avocado.
5. Place the slices of potato into the toaster until potato is a slightly crispy (well done). You can toast until desired doneness.
6. Spread the avocado on potato and season with salt and pepper to taste. Top with pea shoots or microgreens and enjoy.

Nutritional Information:
Calories 210, Fat 13 g, Sodium 115 mg, Carbohydrates 27 g, Fiber 8 g, Sugar 3 g; Protein 4 g.

Tofu Scramble

Yield: Serves 4
Prep Time: 10 minutes
Cook Time: 12 minutes

This vegan-friendly dish is anti-inflammatory. And with the onions, garlic, and mushrooms, it helps the body do its natural job of detoxing. The pickled jalapeños add a nice flavor, and the nutritional yeast adds some B12 and folic acid to your diet.

Ingredients

½ cup sweet onion
2 tablespoons of liquid aminos
1 garlic clove
½ cup of chopped tomatoes
8 oz baby bella mushrooms
1 block of super firm tofu, 16 ounces
4 tablespoons of nutritional yeast
2 teaspoons of turmeric powder
½ teaspoon cumin
2 cups of kale
2 tbsp of pickled jalapeños
Salt and pepper to taste

Directions

1. Place onions and garlic in a preheated pan, adding liquid aminos as needed so they don't stick.
2. Add chopped mushrooms and tomatoes to the pan and sauté for a minute, add aminos as needed.
3. Crumble tofu in your hands and break up into the pan.
4. Add nutritional yeast, turmeric, and cumin and mix in for a minute, adding aminos as needed.

5. Add kale and jalapeños and cook until tender.
6. Add salt and pepper to taste.

Nutritional Information:
Calories 180, Fat 6 g, Sodium 360 mg, Carbohydrates 17 g, Fiber 5 g, Sugar 2 g; Protein 18 g.

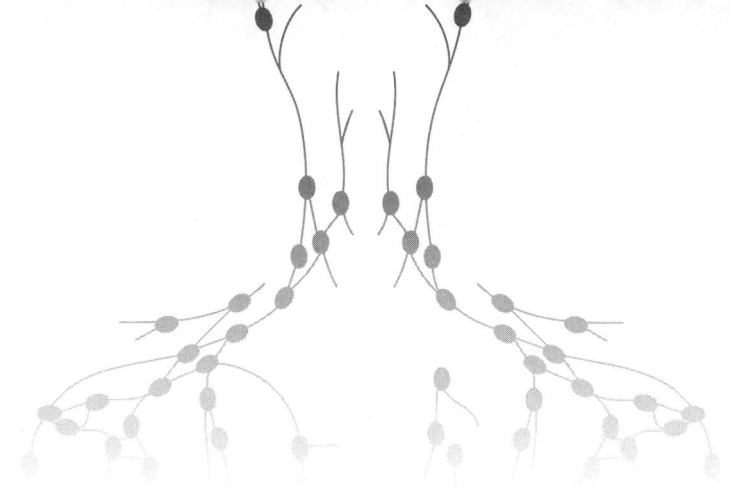

SOUPS AND SALADS

Mushroom and Quinoa Soup

Yield: 6 servings
Serving Size: 1 cup
Prep Time: 5 minutes
Cook Time: 20 minutes

This light soup naturally supports liver and kidney detox. The garlic and lemon zest also support detox, while adding a punch of flavor. The mushrooms create an umami flavor, and the quinoa provides some fiber and protein. This soup is so light that it's great all year round.

Ingredients

½ cup uncooked quinoa
2 tablespoons butter
1 cup onion, chopped
12 ounces cremini mushrooms, washed and sliced
1 clove garlic, minced
1 teaspoon tomato paste
½ teaspoon salt
¼ teaspoon black pepper
4 cups of organic vegetable or chicken broth
1 teaspoon coconut aminos
1 teaspoon lemon zest
1 tablespoon parsley, chopped

Directions

1. In a small pot, add the half cup of quinoa and 1 cup of water. Bring to a boil, then cover and reduce heat to a simmer. Cook until the water is absorbed or for about 20 minutes.
2. While the quinoa is cooking, place butter into a Dutch oven or large pot and melt. Then, add the onions, garlic, and mushrooms.

Cook until mushrooms soften, and onions are translucent; about 2–3 minutes.
3. Add bay leaf, tomato paste, salt, and pepper, stirring once or twice.
4. Slowly, add the 4 cups of broth and 1 teaspoon coconut aminos. Bring to a low boil and then reduce heat, cooking for another 5 minutes.
5. Add teaspoon of lemon zest, stir, and serve hot.

Nutritional Information:
Calories 130, Fat 4 g, Sodium 670 mg, Carbohydrates 20 g, Fiber 3 g, Sugar 4 g; Protein 5 g.

Butternut and Sweet Potato Soup

Yield: 6 servings
Serving Size: 1 cup
Prep Time: 20 minutes
Cook Time: 45 minutes

A delightful soup that provides warmth and coziness; several ingredients support liver and lymphatic detox, from the squash and potatoes to the garlic, rosemary, and even the parsley garnish.

Ingredients

1 1/2 butternut squash, peeled and cubed
2 sweet potatoes, about 8 ounces, peeled and cubed
1 garlic clove, peeled
2 tablespoons olive oil
2 tablespoons butter
1/2 teaspoon salt
1 cup onion, chopped
1/4 cup shallot, chopped
6 cups vegetable broth
1 teaspoon fresh thyme
2 teaspoon fresh rosemary, chopped
Parsley to garnish

Directions

1. Preheat oven to 425°F.
2. Toss the squash, sweet potatoes, and garlic together with 2 tablespoons of olive oil on a sheet pan and cook in the oven for 30 minutes.
3. Meanwhile, prep the onions, shallots, and rosemary.

4. When the vegetables are almost finished cooking, melt the butter in a Dutch oven and add the onions and shallots. Cook for about 5 minutes until the onion is translucent.
5. When the vegetables are finished roasting, place them into the Dutch oven and mix with the onion mixture.
6. Add the 6 cups of vegetable broth, thyme, and rosemary and let simmer.
7. Using an immersion blender or masher, puree the mixture until smooth. Sprinkle parsley on top before serving.

Note: if the soup is too thick for you, add more vegetable broth, a half a cup at a time, until desired consistency is reached.

Nutritional Information:

Calories 360, Fat 9 g, Sodium 870 mg, Carbohydrates 71 g, Fiber 13 g, Sugar 17 g; Protein 7 g.

Roasted Vegetable Caesar

Yield: 5 servings
Prep Time: 15 minutes
Cook Time: 40 minutes

Cruciferous vegetables are researched to be great for metabolic detoxification and liver support. The green crucifers also contain chlorophyll, which enhances detoxification. Cruciferous vegetables such as broccoli, cauliflower, cabbage, and kale contain compounds called glucosinolates, which upregulate the liver's detoxification enzymes. This is a delicious take on a traditional Caesar salad that can be served warm or cold.

Ingredients

4 tsp avocado oil
5 cups cauliflower
3 beets
1 medium red onion
5 cups broccoli
1/2 teaspoon salt, divided
1/4 teaspoon black pepper, divided
1/4 teaspoon garlic powder
10 oz kale
Juice of 1/2 lemon
5 tablespoons sundried tomatoes, packed dry
1/2 cup pumpkin seeds

Directions

1. Preheat the oven to 400°F.
2. Wash, peel, and chop beets to 1-inch pieces.
3. Toss in avocado oil and salt.
4. Place in oven for 15 minutes.

5. While beets are cooking, chop cauliflower and broccoli into bite-size pieces and red onion into slices.
6. Toss in mixing bowl with avocado oil, salt, and pepper.
7. When the 15 minutes are up, place beets back in the oven along with cauliflower, broccoli, and onions for 25 minutes.
8. In the meantime, chop kale into bite-size pieces and massage with lemon juice.
9. Prepare Caesar dressing. (See below.)
10. Let vegetables cool for 5 minutes and then assemble on top of a bed of kale.
11. Top with pumpkin seeds, sundried tomatoes, and Caesar dressing.

Nutritional Information:
Calories 180, Fa 6 g, Sodium 360 mg, Carbohydrates 28 g, Fiber 8 g, Sugar 9 g; Protein 9 g.

Caesar Dressing

Yield: 5 servings
Prep Time: 5 minutes
Cook Time: 3 minutes

The highly-alkalizing hemp seed contains more chlorophyll than most seeds, an ingredient that is great for inflammation and detoxification.

Ingredients

3/4 cup raw cashew pieces
1/4 cup hemp seeds, hulled
3/4 cup water
1 large lemon, juiced
1 1/5 tablespoon nutritional yeast
1 tablespoon stone ground mustard, no salt added
1 tablespoon capers
1/8 teaspoon garlic powder

Directions

1. Measure out all ingredients.
2. Add to the blender and blend for 1–2 minutes.
3. Serve and enjoy!

Nutritional Information:
Calories 190, Fat 14 g, Sodium 95 mg, Carbohydrates 10 g, Fiber 2 g, Sugar 2 g; Protein 9 g.

Cabbage Slaw

Yield: 6 servings
Serving Size: 1 cup
Prep Time: 10 minutes

This is a colorful salad that helps the body detox through different nutrients. Did you know that sesame seeds aid in digestion and liver detoxing? The edamame beans contain isoflavones that help with neutralizing toxins that eventually are excreted. And don't forget that the sulfur in the cabbages also helps with liver detoxing. The light dressing is one I make at home on a regular basis. It's easy and quick for those of us with busy lives.

Ingredients

3 cups red cabbage, thinly shredded
4 cups white cabbage, thinly shredded
2 carrots, grated
1/4 sesame seeds
1 cup edamame beans

Directions

1. Combine red, white cabbage, and carrots in a large bowl.
2. Add edamame beans; mix well.

Nutritional Information:
Calories 180, Fat 19g, Sodium 40mg, Carbohydrates 6g, Fiber 0g, Sugar 2g, Protein 0g

Cabbage Slaw Dressing

Ingredients

1/4 cup apple cider vinegar
1/2 cup extra virgin olive oil
2 tablespoons tbsp raw honey
Salt and pepper to taste

Directions

1. In a bowl, mix all ingredients. This dressing can be stored for three to five days in the refrigerator.

Nutritional Information:
Calories 70, Fat 1.5g, Sodium 40mg, Carbohydrates 11g, Fiber 4g, Sugar 5g, Protein 5g

Herbed Black Bean and Quinoa Salad

Yield: 4–6 servings
Prep Time: 10 minutes
Cook Time: 15 minutes

Black beans are an excellent source of magnesium, which helps the body handle stresses and promotes a healthy nervous system. They also contain selenium, a mineral that aids the liver in detoxing by supporting the liver enzymes. The quinoa is not only high in fiber and protein, but also aids in the detox process.

Ingredients

1/2 cup quinoa, dry
1 teaspoon olive oil
2 garlic cloves, minced
1 bunch swiss chard, destemmed, chopped
1 bunch spinach, destemmed, chopped
1 teaspoon coriander, ground
1 teaspoon paprika, ground
1 1/2 cups dry black beans, cooked OR 1 can organic black beans, low-sodium, rinsed
1 medium cucumber, diced
1 cup cherry tomatoes, halved
1/3 cup tender herbs such as cilantro. dill, mint, parsley, chopped
3 tablespoons tbsp lemon juice
1/3 cup olive oil
Salt and pepper to season
2 tbsp sunflower seeds

Directions

1. In a small saucepan, bring quinoa and 1 cup of water to a boil. Reduce heat to low, cover, and simmer until just about all the water has been absorbed, about 10 minutes.
2. While the quinoa is cooking, add 1 teaspoon of olive oil to a pan along with the garlic and cook until fragrant over medium low heat. Add chopped spinach and swiss chard and cook until wilted. Season with coriander, paprika, salt, and pepper, then add to the cooked quinoa.
3. In a large bowl, combine black beans, cucumber, tomatoes, and herbs. Add quinoa and greens, then mix to combine.
4. In a small bowl, mix lemon juice and olive oil together and season with salt and pepper, then pour over quinoa and bean mixture. Toss to coat and sprinkle with sunflower seeds to finish.
5. This dish can be served either warm or cold, depending on preference.

Nutritional Information:
Calories 180, Fat 6 g, Sodium 360 mg, Carbohydrates 28 g, Fiber 8 g, Sugar 9 g; Protein 9 g.

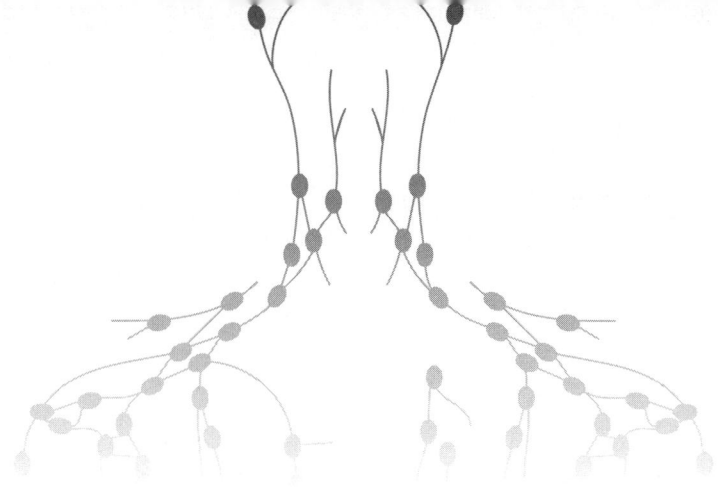

VEGETABLES/SIDES

Sautéed Kale and Toppings
Sautéed Dandelion Greens and Radicchio
Watercress Salad
Roasted Cauliflower with Anchovies
Collard Greens with Apples
Cilantro Hummus
Seedy Crackers

Sautéed Kale and Toppings

Yield: 2 servings
Serving Size: 2 cups
Prep Time: 7 minutes
Cook Time: 5 minutes

This side dish can double as a main meal. The kale contains sulfur and is great for helping the liver. The bonus here is the garlic and onion, which contain allicin for helping with detoxing throughout the lymphatic system and for digestion. If you add the avocado, you are increasing the antioxidants and glutathione that help with inflammation and fighting off toxins. You can get creative and add your own choice of toppings.

Ingredients

1 tablespoon coconut oil
2 cloves garlic, sliced
1/2 large onion, chopped
4 cups of kale, washed and chopped
1/4 cup walnut, chopped

Toppings

1 avocado, cubed
1/2 cup plum tomatoes
Feta cheese (optional)

Directions

1. Heat 12-inch sauté pan. Place coconut oil in pan and allow to melt.
2. Add onions and garlic; cook until onions are soft or translucent.

3. Add chopped kale and chopped walnuts. Cook for about 2 minutes or until kale softens and is bright green.
4. Place on kale mixture on plate and add toppings of your choice.
5. Serve hot.

Nutritional Information:
Calories 370, Fat 29 g, Sodium 60 mg, Carbohydrates 29 g, Fiber 10 g, Sugar 3 g; Protein 10 g.

Sautéed Dandelion Greens and Radicchio

Yield: 2 servings
Serving Size: 1 1/4 cup
Prep Time: 5 minutes
Cook Time: 5 minutes

Nothing says detoxing more than dark leafy greens, and the dandelion in this recipe does the trick. You can always swap out the dandelion for arugula, kale, or mustard greens. Radicchio brings so much nutrition to the dish, starting with detoxing the liver. It also stops the growth of harmful bacteria, promotes the production of bile, and, since it is loaded with so many of the B vitamins, it's great for brain fog.

Ingredients

1 tablespoon olive oil
1/4 red onion, sliced thin
1/2 apple, with skin sliced into
1 garlic clove, minced
1 cup dandelion greens, roughly chopped
1 cup radicchio, roughly chopped
1 teaspoon lemon zest
Salt and pepper to taste

Directions

1. Heat olive oil in sauté pan and add red onion and garlic. Sauté until translucent, about 1 minute.
2. Add the apple and sauté for 30 seconds, then add dandelion and radicchio. Sauté for another minute until wilted.
3. Once greens are wilted, add the lemon zest and sprinkle salt and pepper to taste.

Serving suggestion: double the recipe and add a protein such as tofu or your favorite meat.

Nutritional Information:

Calories 100, Fat 7 g, Sodium 25 mg, Carbohydrates 10 g, Fiber 2 g, Sugar 5 g; Protein 1 g.

Watercress Salad

Yield: 2 servings
Prep Time: 10 minutes

This is a simple but flavorful salad. Watercress increases detoxification enzymes and contains antioxidant compounds, including phenols. It's loaded with nutrients, especially vitamin K and C. This green is an overall powerhouse, nutritionally speaking, and will build up your immune system. You can always serve it with any main meal or add your favorite protein source.

Ingredients

2 cups watercress, washed
2 cups Bibb lettuce, washed and roughly chopped
3 radishes, washed and julienne
2 tablespoons almond slices, divided
3 tablespoons olive oil
¼ teaspoon Dijon mustard
1 tablespoon apple cider vinegar

Directions

1. Mix both greens and radishes in a large bowl.
2. In a separate bowl, pour in vinegar and slowly add the olive oil and stir in mustard. Gently combine.
3. Spilt the greens. Mix into separate bowls, top with almonds, and drizzle the dressing on top.
4. Add salt and pepper to taste.

Nutrition Information:
Calories 310, Fat 24 g, Sodium 135 mg, Carbohydrates 23 g, Fiber 9 g, Sugar 13 g; Protein 5 g.

Roasted Cauliflower with Anchovies

Yield: 2 servings
Prep Time: 10 minutes
Cook Time: 40 minutes

Ingredients

3 teaspoons of olive oil
3 anchovy fillets, drained
1/8 teaspoon crushed red pepper
4 cups cauliflower florets
1/2 + 1/8 teaspoon turmeric
1/2 teaspoon lemon zest

Directions

1. Preheat oven to 450°F.
2. In a sauté pan over medium to high heat, add oil, then add the anchovies, crushed red pepper, and turmeric and cook about 3 minutes or until the anchovies break down.
3. Add the cauliflower florets to the pan and mix with the anchovy and oil mixture. Place on a baking sheet and pour the anchovy mixture over the cauliflower and toss to mix well. Bake at 450°F for about 25 minutes until cauliflower is tender.
4. Sprinkle with lemon zest, toss, and serve.

Nutrition Information:
Calories 250, Fat 22 g, Sodium 280 mg, Carbohydrates 12 g, Fiber 5 g, Sugar 4 g; Protein 6 g.

Collard Greens with Apples

Yield: 2 servings
Serving Size: 2 cups
Prep Time: 5 minutes
Cook Time: 5 minutes

Collard greens are loaded with calcium and vitamin D and aid in detoxing. The apple adds more fiber and takes a little bit of the bitterness away from the greens.

Ingredients

2 tablespoons olive oil
1 clove garlic, sliced
1/2 teaspoon turmeric, grated
5 cups collard greens, washed and sliced
1 McIntosh apple, sliced, skin on
Salt and pepper to taste

Directions

1. Place a sauté pan over medium heat; add oil, then add garlic and turmeric. Cook for about 30 seconds until the garlic turns opaque but not brown.
2. Add collard greens and stir, cooking for another minute, then add apples.
3. Cook mixture with apples for about 1 to 2 more minutes or until the apples start to soften.
4. Add salt and pepper to taste and serve.

Nutrition Information:
Calories 200, Fat 15 g, Sodium 15 mg, Carbohydrates 19 g, Fiber 6 g, Sugar 10 g; Protein 3 g.

Cilantro Hummus

Yield: 2 servings
Serving Size: 2 tablespoons per serving, 16 servings per 2 cups
Total Time: 5 minutes

Chickpeas and cilantro make this recipe a double header for detoxing. The peas are high in fiber and contain protein, while the cilantro is high in antioxidants and improves the immune response. Garlic, tahini, and lemon juice add to the detoxing properties of this tasty recipe. Serve with jicama or celery, or use as a spread.

Ingredients

1 medium garlic clove
16 ounces chickpeas, washed and rinsed
1 bunch cilantro (about 2 ounces) stems included
1/4 cup olive oil
1 1/2 tablespoons fresh lemon juice
2 tablespoons tahini
1/2 teaspoon salt
1 ounce water

Directions

1. Place the garlic in a food processor and pulse.
2. Add chickpeas and cilantro, then pulse.
3. While the food processor is running, add the oil and lemon juice, then add tahini and blend.
4. Add salt and drizzle in the water slowly while the blender is running.
5. Store in the refrigerator for up to 5 days.

Nutrition Information:
Calories 70, Fat 5 g, Sodium 160 mg, Carbohydrates 5 g, Fiber 0 g, Sugar 1 g; Protein 2 g.

Seedy Crackers

Yield: 24 crackers
Prep Time: 4 minutes
Cook Time: 2–2 ½ hours

Seeds provide Omega-3 fatty acids, helping the microbes in your gut microbiome cleanse themselves. In turn, this process lowers inflammation since the microbiome is closely linked with the immune system. These seeds are also packed with magnesium to help support the nervous system. Feel free to add in different spices to create different flavored crackers. They pair wonderfully with hummus and other dips.

Ingredients

1/4 cup chia seeds
1/2 cup flax seeds
1/2 cup pumpkin (pepita) seeds
1/4 cup sunflower seeds
1/4 cup sesame seeds
1/2 teaspoon salt
1 tablespoon fresh thyme leaves
1 teaspoon garlic powder
1/2 teaspoon coriander, ground
3/4 cup water

Directions

1. Preheat oven to 200°F and line a baking sheet with parchment paper.
2. In the bowl of a food processor, add seeds, salt, thyme, garlic powder, and coriander. Blend until fully incorporated, and the pumpkin seeds are about the same size as the smaller seeds.
3. Add seed mixture to a bowl and add water. Mix for 1–2 minutes until the seeds begin to gelatinize and thicken.

4. Spread mixture onto the parchment paper in a thin layer. With a butter knife, score the sheet into 20–24 crackers. Note: it does not have to be perfect.
5. Dehydrate the crackers in the oven for 2 to 2 1/2 hours until there is little to no moisture left.
6. Store in an airtight container for up to a week.

Nutritional Information:
Calories 1280, Fat 94 g, Sodium 1450 mg, Carbohydrates 59 g, Fiber 44 g, Sugar 1 g; Protein 54 g.

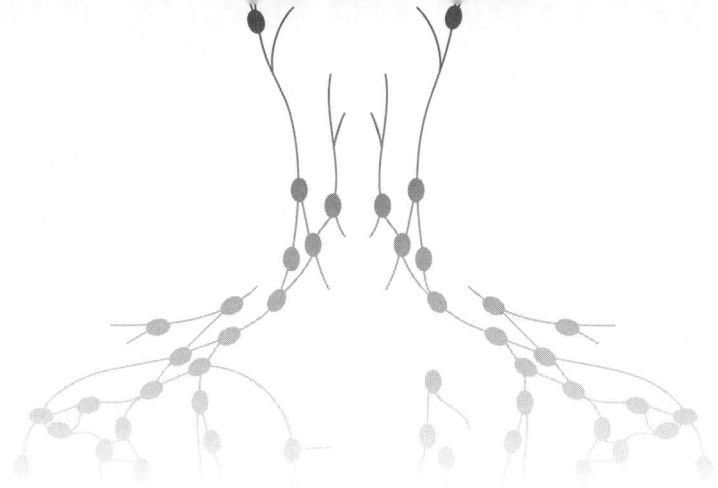

MAIN MEALS

Apple Chicken
Sauteed Sardines with Fennel
Lamb with Rosemary
Chickpeas and Spinach
A Black Lentil Bowl
Steamed Halibut
Cauliflower Rice Bowl
Lentil Mushroom "Meatballs" with Romesco Sauce
Romesco Sauce
Wild Salmon with White Beans and Greens

Apple Chicken

Yield: 2 servings
Serving Size: 4 ounces
Prep Time: 5 minutes
Cook Time: 15 minutes

This is a quick dish to prepare for any night of the week. Serve it with steamed asparagus for an extra detox. The spinach and mushrooms will help with detoxing, while the chicken provides protein and will help with satiety.

Ingredients

2–4 ounces chicken breast, thinly sliced
1 ounce olive oil
¼ cup Macintosh apple, chopped, skin on
2 ounces fresh spinach
¼ cup shitake mushrooms, chopped
½ teaspoon thyme
Salt and pepper to taste
1 teaspoon grainy Dijon mustard
¾ cup chicken broth, split

Directions

1. Preheat oven to 350°F.
2. Heat olive oil in a sauté pan, add apples, and sauté for about 2 minutes or until softened. Once the apples are soft, add the mushrooms, spinach, and thyme. Cook for about 1 minute or until the spinach has wilted.
3. Take the apple-spinach mixture and divide, spreading evenly over two chicken breasts.
4. Sprinkle with salt and pepper to taste.

5. With two hands, take the thinner end of the chicken breast and roll it up tight. You can use cooking twine to tie up the rolls or keep them together with toothpicks.
6. Place the rolls in a small baking dish and spread the Dijon mustard evenly over the two breasts. Pour 1/2 cup of chicken broth in the bottom of the pan and place in the oven. Cook for 15–20 minutes or until the chicken reaches 165°F.
7. Once the chicken is cooked, scrape the mustard off the chicken into the pan. Then, remove the chicken breasts from the pan and carefully place them on the cutting board.
8. Take a small roasting pan and place on the stovetop over low heat. Add the rest of the chicken broth and gently stir until mixed well.
9. Slice the chicken breasts width-wise to create pinwheels. Once plated, drizzle the sauce over the pinwheel slices.

Nutrition Information:
Calories 330, Fat 19 g, Sodium 1100 mg, Carbohydrates 5 g, Fiber 1 g, Sugar 2 g; Protein 37 g.

Sauteed Sardines with Fennel

Yield: 1 serving
Prep Time: 5 minutes
Cook Time: 5 minutes

This Moroccan-inspired dish is light and full of flavor. The lemon adds a nice zest as well as contributing the detoxing powers of the sardines. Sardines are high in Omega-3, a good type of fat that helps with liver detoxing and lowering inflammation. Omega-3 also builds up the immune system. This would be great served on a bed of arugula. Double the recipe ingredients if you are cooking for two.

Ingredients

1 teaspoon olive oil
1 garlic clove, sliced thin
1 1/2 cup fennel, sliced very thin
1 ounce shitake mushrooms, sliced
1 4.37 ounce can of wild sardines, packed in oil
15 cherry tomatoes, cut in half
2 tablespoons lemon juice, fresh squeezed
1/2 teaspoon thyme

Directions

1. Heat olive oil in a sauté pan over medium heat, then add garlic and sauté for 30 seconds.
2. Add fennel and mushrooms and sauté until soft, about 1 to 2 minutes
3. Add the sardines and heat through.

4. Add tomatoes, lemon juice, and thyme, then stir, making sure not to break up the sardines.
5. Serve hot over a bed of arugula.

Nutritional Information:

Calories 150, Fat 6 g, Sodium 85 mg, Carbohydrates 25 g, Fiber 8 g, Sugar 13 g; Protein 5 g.

Lamb with Rosemary

Yield: 2 servings
Serving Size: 2 2-ounce chops
Prep Time: 5 minutes
Cook Time:12 minutes

If you are an omnivore, this lamb dish might be a choice to consider when detoxing. The combination of the lemon, garlic, and rosemary make it a potpourri of medicinal powers. The lemon has enzymes and vitamin C that support digestion and liver detox. Rosemary helps with the inflammatory response in the liver and promotes brain synapses. Serve with a side of sautéed kale.

Ingredients for Marinade

4 tablespoons olive oil
2 tablespoons lemon juice
1 teaspoon fresh rosemary
1 large garlic clove, minced
1/2 teaspoon salt
1/4 teaspoon pepper
4 2-ounce lamb chops (2 chops per person)

Directions

1. Mix olive oil, lemon, rosemary, garlic, salt, and pepper.
2. Place lamb in bowl and toss with marinade. Let sit for at least 15–20 minutes.
3. Heat pan and add the lamb chops. Cook lamb on one side and turn over when brown. Finish cooking lamb until it reaches 145°F.

4. Serve immediately and serve with quinoa salad or grilled asparagus.

Nutritional Information:
Calories 480, Fat 45 g, Sodium 620 mg, Carbohydrates 2 g, Fiber 0 g, Sugar 0 g; Protein 19 g.

Chickpeas and Spinach

Yield: Serves 2; 1 1/4 cup preserving
Prep Time: 5 minutes
Cook Time: 10 minutes

This Indian-inspired dish is great alone or served over quinoa or basmati rice. It has all the elements of a fall or winter dish. Although it's vegan, there is protein in the chickpeas—and more if you serve the dish over quinoa. Detoxing is no surprise here with inclusion of ginger, spinach, fresh lemon. Double the recipe and have some for leftovers.

Ingredients

1 tablespoon coconut oil
1/4 cup onions, chopped
2 teaspoons fresh ginger, minced
1 tablespoon curry powder
2 cups cooked chickpeas
1 cup unsweetened coconut milk
Salt and pepper to taste
2 cups fresh spinach, packed
Lemon rind from 1/2 a large lemon
1 tablespoon fresh lemon juice

Directions

1. Heat coconut oil in a medium pot, add onions and ginger, and sauté until onions are translucent, about 1 minute.
2. Sprinkle in curry powder, stirring until fragrant.
3. Add the cooked chickpeas, stirring once or twice, then pour in coconut milk. Heat through for about 1 more minute.
4. Add in 2 cups of spinach and allow the spinach to wilt.

5. Add the lemon juice and lemon rind, stir once more, and season with salt and pepper.
6. Serve alone or over quinoa.

Nutritional Information:
Calories 400, Fat 14 g, Sodium 560g, Carbohydrates 55 g, Fiber 15 g, Sugar 12 g; Protein 16 g.

A Black Lentil Bowl

Yield: 3 servings
Serving Size: 2 cups
Prep Time: 10 minutes
Cook Time: 25 minutes

This lentil bowl is high in fiber and helps with digestion and constipation. Lentils also provide a natural way of helping the liver detox. In addition, they are a great source of protein and iron. So, whether you are vegan or not, this is an all-around winner. I love the subtle hint of cardamom that lingers after you have your first bite. This spice contains antioxidants and may be helpful in lowering blood pressure. It is one of my favorite spices for its unique flavor.

Ingredients

3/4 cup onions, chopped
1 large garlic clove, minced
1 teaspoon ginger, minced
1 tablespoon olive oil
1/2 teaspoon cumin
1/2 teaspoon curry powder
1/2 teaspoon ground cardamom
3/4 cup black lentils
1 cup carrots, diced
½ cup unsweetened coconut milk
2 cups vegetable broth
14.5-ounce can diced tomato, low sodium
1 1/2 cups cauliflower, cut into small bite size pieces
3 tablespoons cilantro, chopped for garnish

Directions

1. In a Dutch oven, heat olive oil, then add onions, garlic, and ginger. Sauté until onions are translucent, about 1 minute.
2. Add cumin, curry powder, and cardamom to the pot, then stir until fragrant, about 1 minute.
3. Next, add lentils, carrots, broth, coconut milk, and tomatoes. Stir until combined.
4. Season with salt and pepper to taste.
5. Add cauliflower, bring to a boil, then return to a simmer. Cook until vegetables and lentils are tender, about 20 more minutes.
6. Serve in a bowl or over rice or quinoa and garnish with cilantro.

Nutritional Information:
Calories 320, Fat 7 g, Sodium 510 mg, Carbohydrates 52 g, Fiber 14 g, Sugar 10 g; Protein 17 g.

Steamed Halibut

Yield: 2 servings
Serving Size: 2 4-ounce pieces
Prep Time: 5 minutes
Cook Time: 10 minutes

This light dish is a detox powerhouse, with many ingredients supporting the immune system and aiding in digestion. The hint of coconut aminos makes the flavors come together. It's a light and airy meal that I suggest serving with kale for an added detoxifying punch. You can also serve it with cabbage slaw.

Ingredients

1 1/2-ounce coconut milk
1 tablespoon coconut aminos
1 teaspoon ginger, grated
1 1/2 teaspoon fresh lime juice
8-ounce halibut
Salt and pepper to taste
1 teaspoon cilantro
2 shitake mushrooms, sliced (optional)

Directions

1. Preheat oven to 400°F.
2. Mix coconut milk, coconut aminos, grated ginger, and lime juice in a bowl and stir well.
3. Place the halibut in a baking dish that has a matching cover or place the fish on a piece of parchment paper and then on an oven-proof pan.
4. Pat the halibut dry and sprinkle salt and pepper over the fish.
5. Pour the coconut milk mixture over the fish and top with cilantro and mushrooms.

6. Fold the parchment paper over to close it up with the fish in it. If you are using a baking dish, put the lid on to cover the fish and mixture.
7. Cook for 10 minutes.

Nutritional Information:

Calories 190, Fat 6 g, Sodium 260 mg, Carbohydrates 6 g, Fiber 1 g, Sugar 4 g; Protein 26 g.

Cauliflower Rice Bowl

Yield: 2 servings
Prep Time: 8 minutes
Cook Time: 10 minutes

This dish is great for helping with detoxing the liver, thanks to the sulfur-containing vegetables, cauliflower, and bok choy. Although it is simple, it's flavorful with the addition of coconut aminos and sesame oil. I love this dish on its own or with pan-seared tofu or leftover chicken to up the protein.

Ingredients

1 1/2 teaspoons olive oil
2 cups riced cauliflower
2 tablespoons bok choy, washed and sliced thin (include some of the green parts too)
2 tablespoons coconut aminos
2 tablespoons water
1 cup shitake mushrooms, sliced
1 tablespoon sesame oil
1 tablespoon scallions

Directions

1. In a large pan, heat oil over medium heat. Add cauliflower and cook over high heat for about 2 minutes or until heated through, stirring constantly.
2. Still on high heat, add the bok choy and cook for another 2 more minutes or until the bok choy softens slightly.

3. Add mushrooms and continue stirring. Add the coconut aminos and water to the pan and continue to stir.
4. Add scallions and sesame oil and stir to mix in with vegetables.

Nutritional Information:
Calories 150, Fat 10 g, Sodium 510 mg, Carbohydrates 14 g, Fiber 3 g, Sugar 9 g; Protein 2 g.

Lentil Mushroom "Meatballs" with Romesco Sauce

Yield: 16 "meatballs"
Prep Time: 10 minutes
Cook Time: 1 hour

This dish packs a punch for detoxing with many of the ingredients, including the golden lentils, mushrooms, garlic, onion, and the mixture of spices. The mushrooms help with inflammation as well, making it easier for the body to perform at its optimal level.

Ingredients

3 red bell peppers
1 cup golden lentils, dry, rinsed
1 tablespoon olive oil
1 cup white onion, diced
2 cloves garlic
4 ounces shiitake mushrooms
2 medium carrots, peeled, diced
3 ribs celery, diced
1/4 teaspoon salt
1/8 teaspoon pepper
1/2 cup tender chopped herbs such as basil, tarragon, cilantro, parsley, or dill.

Directions

1. Preheat oven to 350°F and line a baking sheet with parchment paper.
2. In a small saucepan, add lentils and cover with water. Bring to a boil and then reduce heat to low and simmer until al dente, about 10–12 minutes. Strain and set aside.

3. In a large pan over medium heat, add olive oil, onion, and garlic. Sautee until fragrant, about 2–3 minutes and then add mushrooms, carrots, and celery. Mix to combine and cook 3–4 more minutes. Season with salt and pepper and cook until the carrots and celery are soft.
4. Remove pan from heat and mix in tender herbs. Add mixture into the bowl of a food processor along with the cooked lentils. Blend until fully incorporated.
5. Shape mixture into 16 balls and place on lined baking tray. Bake for 30–35 minutes, flipping hallway through. Serve with romesco sauce. (See below.)

Nutritional Information:
Calories 280, Fat 5 g, Sodium 220 mg, Carbohydrates 478 g, Fiber 10 g, Sugar 9 g; Protein 16 g.

Romesco Sauce

Yield: 6–8 servings
Prep Time: 10 minutes
Cook Time: 30 minutes

Ingredients

3 red bell peppers
1 cup cherry tomatoes
1 garlic clove
1 cup almonds, whole
1 tablespoon smoked paprika
1/4 teaspoon salt

Directions

1. Set oven to broil on high. Place peppers on a baking tray and put under the broiler, flipping every minute or two to char all sides.
2. Once peppers are charred and softened, place in a bowl and cover with a lid to trap the steam. Set aside to cool, about 15–20 minutes.
3. Once cool, remove the skin, stem, and seeds of the roasted peppers. Place into blender and add cherry tomatoes, garlic, almonds, smoked paprika, and salt. Blend until smooth.
4. Taste and adjust seasoning as needed. Serve over Lentil Mushroom "Meatballs" and pasta or noodles of choice.

Nutritional Information:
Calories 130, Fat 9 g, Sodium 75 mg, Carbohydrates 8 g, Fiber 4 g, Sugar 4 g; Protein 5 g.

Wild Salmon with White Beans and Greens

Yield: 2 servings
Serving Size: 4 ounces
Prep Time: 5 minutes
Cook Time: 15–20 minutes

In this recipe, the Omega-3s from the salmon together with the greens is a double-detox whammy. Note: the darker the greens, the more antioxidants they contain. Double the recipe and have it for lunch the next day.

Ingredients

1 tablespoon olive oil
1/4 cup fresh lemon juice
2 4-ounce salmon fillets
2 garlic cloves, minced and divided
1 teaspoon thyme
2 cups white beans, cooked
2 cups dark greens, your choice—mustard greens, dandelion greens, kale, watercress, or arugula
Salt and pepper

Directions

1. Preheat oven to 400°F and line a small baking sheet with parchment paper.
2. Place oil, lemon juice, 1 minced garlic clove, salt, and pepper in a shallow bowl. Dip the salmon into the bowl, coating it on each side.
3. In a sauté pan over medium heat, place the salmon skin side up into the pan and cook for about 2 minutes or until salmon develops a nice golden color.
4. Place the seared salmon on the prepared baking sheet and bake for about 12–15 minutes or until the salmon flakes easily with

a fork. (Cooking may vary depending on the thickness of the fillet.)
5. While the salmon is baking, using the same pan used for the salmon, sauté the garlic over medium heat for 30 seconds, add thyme, and cook for 10 seconds until heated through.
6. Add the white beans and greens of your choice. Cook until greens are wilted, and beans are heated.
7. Place the bean and green mixture on a plate and top with the cooked salmon. Serve with watercress soup.

Nutritional Information:
Calories 540, Fat 23 g, Sodium 1720 mg, Carbohydrates 45 g, Fiber 2 g, Sugar 6 g; Protein 38 g.

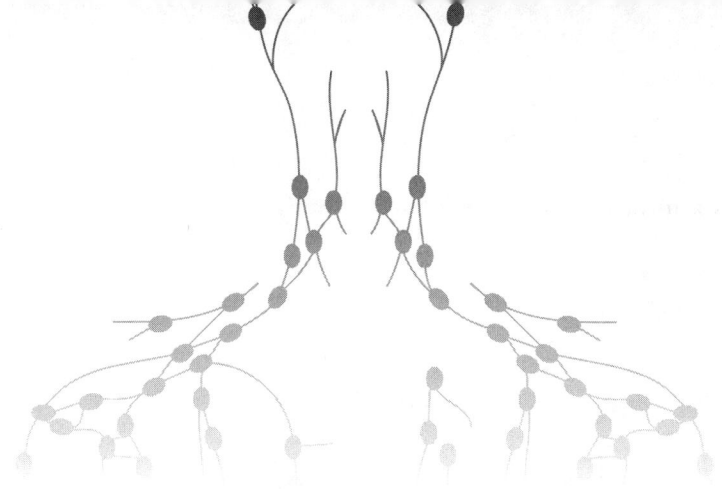

SWEET TREATS

Hemp Butter Peppercorn Cookies
Cranberry Ginger Nice Cream
Banana Split for Two

DR. LORETTA T. FRIEDMAN

Hemp Butter Peppercorn Cookies

Yield: 24 cookies
Time: 28–30 minutes

With this treat, you'll be satisfying your sweet tooth, while boosting your detoxing abilities with the hemp hearts and peppers. It's also gluten free and has no preservatives.

Ingredients

1/2 cup hemp hearts, hulled
1/2 cup honey, raw
1/2 cup almond flour, fine
1/4 teaspoon salt
1/2 teaspoon baking soda
1/2 teaspoon black peppercorns, whole
1/2 teaspoon pink peppercorns, whole
1 teaspoon coriander seeds, whole
1/3 cup apricots, dried, whole, unsulfured, diced

Directions

1. Preheat oven to 325°F and line a baking sheet with unbleached parchment paper.
2. Add hemp hearts to a food processor and blend until a paste is formed, about 1–2 minutes.
3. Add honey, salt, and baking soda, then blend until fully incorporated.
4. Remove mixture from the bowl of the food processor, add to a large bowl, and combine with almond flour.
5. In a small, dry pan, add both peppercorns and coriander seeds and toast spices over low heat until fragrant, about 2–3 minutes.
6. Once toasted, grind spices in a spice grinder or with a mortar and pestle until coarsely ground.

7. Add spices to dough along with diced, dried apricots and mix thoroughly.
8. Divide dough into 24 balls and place on parchment-lined baking sheet, slightly pressing each ball down to form a cookie. You may want to wet your hands a bit to prevent the dough from sticking to your fingers when pressing down.
9. Bake cookies for 8–10 minutes until cookies are slightly golden brown around the edges. Allow cookies to cool on baking sheet for 10 minutes before removing and enjoying.

Nutritional Information:
Calories 60, Fat 2.5 g, Sodium 50 mg, Carbohydrates 8 g, Fiber 1 g, Sugar 7 g; Protein 2 g.

Cranberry Ginger Nice Cream

Yield: 4 servings
Prep Time: 2 minutes
Cook Time: 5 minutes

Cranberries are excellent berries capable of detoxifying the body by increasing healthy lymph flow. Paired together with citrus and ginger, this sweet yet tangy treat helps clear the body of toxins while satisfying your sweet tooth.

Ingredients

4 medium bananas, frozen
1 cup cranberries, fresh or frozen
1 teaspoon orange zest
½-inch piece of fresh ginger, grated
2 tablespoons raw honey

Directions

1. In a small saucepan, add cranberries, orange zest, ginger, honey, and a splash of water and simmer over low heat until cranberries begin to break down and the mixture becomes thickened.
2. In a food processor, add frozen bananas and blend until smooth.
3. Add cranberry mixture to the bananas and pulse just until cranberries are swirled throughout.
4. Place in a freezer-safe container and freeze for 1-2 hours or to desired consistency.

Nutritional Information:
Calories 150, Fat 0 g, Sodium 0 mg, Carbohydrates 39 g, Fiber 4 g, Sugar 24 g, Protein 1 g.

Banana Split for Two

Yield: 2 servings
Prep Time: 8 hours
Cook Time: 10 minutes

Life is about balance. Here is a sweet treat to make you feel like you have indulged, yet the ingredients are real foods and contain no preservatives or added sugars.

Ingredients

2 frozen medium bananas
1 can full-fat coconut cream
1 fresh medium banana
4 medjool dates
1/4 cup hot water
2 tablespoons of pure dark chocolate chips
1/2 cup of canned coconut milk
1 tablespoon crushed walnuts
4 fresh cherries

Directions

1. Peel and freeze two bananas overnight.
2. Refrigerate can of coconut cream overnight.
3. Peel one raw banana and add to bottom of ice cream dish.
4. Add medjool dates and chocolate chips to small blender with ¼ cup of hot water and soak for 5 minutes.
5. Add 2 frozen bananas to food processor with coconut milk and blend on high for 2 minutes.
6. Scoop ice cream out and add to ice cream dish.
7. Blend dates, water, and dark chocolate chips in a high-speed blender for 1 min and pour on top of ice cream.

8. Scoop out coconut cream solids to a bowl and whip with a hand mixer for 3 mins until light and fluffy.
9. Add a scoop on top of ice cream.
10. Add walnuts and cherries and enjoy!

Nutritional Information:

Calories 360, Fat 11 g, Sodium 30 mg, Carbohydrates 68 g, Fiber 8 g, Sugar 43 g; Protein 5 g.

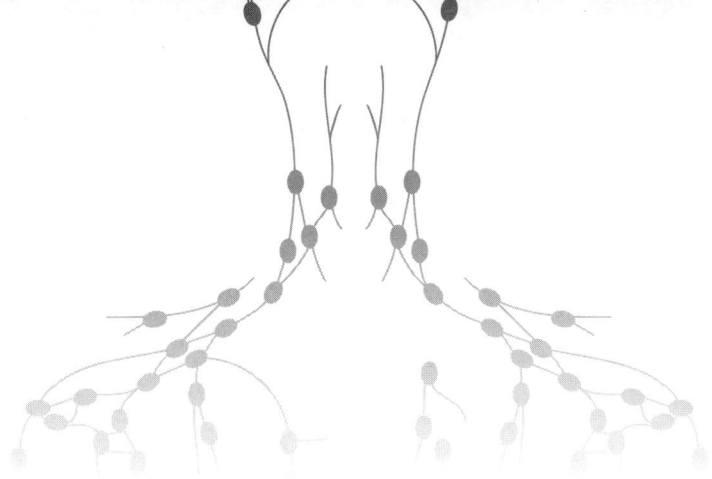

APPENDIX: WHAT YOU NEED TO KNOW

Chapter 1

- Until a person's immune system is challenged, toxicity can go completely undetected.
- COVID-19 long haulers most likely have high toxicity levels that are keeping them from proper recovery.
- The lymphatic system is a vital part of your body's immune system.
- The lymphatic system is an intricate network of vessels through which debris travels via lymph fluid.
- Lymph nodes are bean-shaped structures where this debris gets dumped.
- Your body contains one hundred and sixty lymph nodes in the head and neck region, three hundred in the trunk region, and hundreds more in the lower extremities.
- Toxins can inflame the lymph nodes, causing them to swell, harden, or become extremely tender.
- Lumpy, tender breasts are a sign of toxicity.
- Since most doctors have training only in a specific area, they have limited ability to diagnose what's really happening in your lymphatic system.

- Natural remedies such as turmeric, ashwagandha, and ozone therapy can be highly effective in treating inflammation, compromised adrenal glands, and lymphoma. These are almost never prescribed by western doctors.
- Lymphedema is often the first sign that your body contains high toxicity levels.

Chapter 2

- The liver's main function in the human body is to detoxify substances.
- In order to get to the bottom of your health issues, you must be willing to do some detective work and invest time and money into holistic medical care.
- Don't just take one doctor's word for it. Get several opinions and ask questions.
- Dietary fat can become a toxin and interfere with liver function.
- Alternative treatments such as Ozone therapy and Essiac tea can be highly effective in treating cancer patients.

Chapter 3

- As cancer survivorship improves, long-term side effect risk, particularly for lymphedema, also increases, with 5 to 50 percent of cancer survivors developing lymphedema.
- The medical establishment has widely ignored the extent to which pain often accompanies lymphedema.
- Many common non-steroidal anti-inflammatory drugs (NSAIDs) not only offer minimal pain relief, but also pose new, potentially fatal health problems.
- Up to 29 percent of people who are prescribed opioids for chronic pain end up misusing them.
- Over one hundred million Americans continue to suffer from mistreated chronic pain each year.

- Most western physicians are either uninformed about effective alternative treatments, have conflicts of interest with these treatments, or are fearful of having their practices challenged by pharmaceutical and insurance companies.
- Popular, effective alternative treatments include cannabis, kratom, Omega-3 supplements, low-level laser, and lymphatic drainage.
- Many chronic pain sufferers get stuck in their stories of pain that keep them from seeking the help they need.

Chapter 4

- Directional Non-Force Technique (DNFT) is the only chiropractic technique to fix low back pain in six visits.
- DNFT works directly with the spinal discs to get to the root of the problem.
- Doctors' use of blanket diagnoses to describe a wide range of ailments is arguably a form of malpractice that leaves patients misguided and uninformed.
- The main culprit besides stress for chronic inflammation is a high toxicity level in the body.

Chapter 5

- The number of deaths caused by the 2020 COVID-19 pandemic revealed the high toxicity levels of many Americans.
- The term *lymphedema* allows doctors to ignore patients' underlying toxicity that causes their swelling and inflammation.
- Our water supply contains PCBs, pesticides, fluoride, and other heavy metals that our bodies are absorbing each day.
- Trichloroethylene (TCE), formaldehyde, phenols, and other Volatile Organic Compounds (VOCs) are all common toxins found in our soil, groundwater, and indoor air spaces.

- These chemicals can cause many harmful side effects such as nerve damage, irregular heartbeats, liver and kidney damage, coma, possible kidney cancer, liver cancer, malignant lymphoma, asthma, rare nose and throat cancers, irregular breathing, muscle weakness and tremors, loss of coordination, convulsions, and respiratory arrest.
- Pesticide and herbicide production has increased at a rate of about 11 percent per year, from 0.2 million tons in the 1950s to more than five million tons in 2000.
- Exposure to pesticides and herbicides can result in elevated cancer risks, a disruption of the body's reproductive, immune, endocrine, and nervous systems, and respiratory pathologies such as asthma, chronic obstructive pulmonary disease (COPD), and lung cancer.
- BPA is a synthetic estrogen found in many rigid plastic products, including most water bottles and food packaging.
- BPA throws off the body's hormonal balance and causes hormone-receptor-positive breast cancer.
- You can limit your exposure to Bisphenol acid (BPA) by carrying your own glass, steel, or ceramic water bottle, not cooking food in plastic containers, paying attention to the recycling numbers on plastics, and not reusing plastics that contain the recycling number 1.

Chapter 6

- Regular exercise helps circulate the extracellular fluid that carries toxins away to the liver and kidneys for disposal.
- Exercise also reduces inflammation and clears lymphatic congestion.
- Extra fat cells block cell receptors, making it harder to get rid of toxins.
- Eating whole, unprocessed, organic foods greatly reduces the number of toxins we consume each day.

- Water is also essential to the kidneys' ability to filter and remove waste products from the blood, eliminate toxic substances in the urine, and receive water-soluble toxins from the liver for processing.
- A good amount of water to drink every day is eight to ten glasses.
- Biotoxicity caused by medications can create a vicious cycle by increasing the presence of free radicals and creating hormonal imbalances that might require even more medications.
- Make sure you keep a careful record of your medications and dosages and share it with all of your physicians.

Chapter 7

- Instead of treating toxicity, most doctors employ modalities that provide patients with only short-term relief from symptoms.
- Lymphatic massage alone cannot remove the toxins in your body, nor can it open the blocked ducts of your lymphatic system.
- Lymphedema occurs when toxins do not allow the lymph fluid to drain from internal tissue.
- Lymph-Biologics™ is the only proven technique that actually resolves inflammation and swelling of the lymph nodes.
- Lymph-Biologics™ helps rid the body of toxins by increasing fluid volume and exchange at the site of the lymphatic vessels.
- The question to ask isn't *Are you toxic?* but *How toxic are you?*
- Shrinking cell membranes and hindered cell receptors are by-products of high toxicity levels.
- Detoxification methods of healing have been used for thousands of years.
- Metabolic detoxification helps patients suffering from allergies, anxiety, arthritis, diabetes, headaches, heart disease, high cholesterol, digestive disorders, and more.

Chapter 8

- There is a direct correlation between cancer and the amount of toxins in your system.
- Natural remedies such as Essiac tea and sessions inside an O3-infused "sweat box" help release the carcinogenic toxins out the body's pores.
- The cell test I perform on patients is similar to the bioelectrical impedance analysis (BIA) physicians use to assess body composition.
- Omega-3 is essential to running important reactions in the body, including cognitive functions and COX-2 reaction, which reduces inflammation.
- "Superfoods" such as avocado and coconut oil can be hard on the pancreas and gallbladder when consumed in large amounts and can contribute to back pain and digestive issues.
- Stress increases cortisol levels and throws your body's physiological responses completely out of whack.
- Stress elevates glucose and insulin levels, exhausting the pancreas, the organ that produces insulin.
- Elevated insulin levels also cause weight gain, a telltale sign that a person is stressed.
- Raised cholesterol levels cause the blood vessels to narrow, restricting blood flow and instigating coronary artery disease.
- The Fat Skinny People of the World are starving themselves of the protein needed to support a healthy muscle-to-fat ratio.
- A cell test will help determine environmental toxins, chemical sensitivities, medications, and a person's natural predispositions, factors that all play a role in a person's toxicity level.

Chapter 9

- Premenstrual Syndrome (PMS) is another fake diagnosis. The real problem is estrogen dominance, a condition that is not normal, even though it is common.
- Chronic stress, poor gut and liver health, and environmental toxins can cause this hormonal imbalance.
- Signs of estrogen dominance include bloating, decreased sex drive, tender, lumpy breasts, mood swings, and weight gain.
- Drinking less alcohol, taking care of the bacterial balance in your gut, boosting your fiber intake, and eating organic and hormone-free are all ways to correct estrogen dominance.
- Leading causes of breast cancer include the average American diet, inflammatory conditions, radiation exposure, and especially environmental endocrine disruptors, such as fossil fuel by-products, toxic chemicals, and environmental pollutants.
- Based on US mortality statistics, screening mammography prevents only one death per 1,000 women screened.
- If you are a woman with dense, fibrous breasts, regular mammography might do very little to protect you.
- A woman who has already developed breast cancer risks rupturing the encapsulation of a cancerous tumor when she gets a mammogram.
- Unlike mammography, thermography works on dense breast tissue and doesn't expose women to harmful radiation.
- Angiogenesis, the increased blood supply to cancer cells, produces measurable amounts of heat that can be detected by thermometric devices.
- Women with dense breast tissue and women under forty are overlooked when it comes to screening for breast cancer.
- Thermograms can identify cancer sooner with more accurate results.
- There are many ways to take care of your breasts and lymphatic system, including wearing less constricting bras without

- underwire, dry brushing, taking alternating hot and cold water showers, taking brisk walks, moving your body regularly, drinking a lot of water, gently jumping on a rebounder, massaging your breasts, drinking certain herbal teas, and using paraben- and aluminum-free deodorant.
- Root canal and other dental work can spread toxins and cause inflammation throughout the whole body.

Chapter 10

- Only about one in ten adults meets the federal recommendation of one and a half cups of fruits and two to two and a half cups of vegetables each day.
- A Mediterranean diet, rich in fruits, vegetables, whole grains, nuts, legumes, and olive oil, is known to reduce inflammation by supporting your GI tract.
- I advise my patients to keep their sugar intake to twenty to thirty grams or lower a day.
- The best detoxifying/anti-inflammatory superfoods include citrus, berries, avocados, sunflower seeds, hemp seeds, flax seeds, and greens.
- Lemon water helps fight cancer and alkalinizes the blood.
- Turmeric has amazing anti-inflammatory properties that have been recognized for years.
- Dairy, farm-raised seafood, and caffeine are among the foods you should avoid when trying to detoxify and balance your immune system.